£32

Management for Professionals

For further volumes:
http://www.springer.com/series/10101

Roman Boutellier · Mareike Heinzen

Growth Through Innovation

Managing the Technology-Driven Enterprise

Roman Boutellier
Mareike Heinzen
D-MTEC
ETH Zürich
Zürich
Switzerland

ISSN 2192-8096 ISSN 2192-810X (electronic)
ISBN 978-3-319-04015-8 ISBN 978-3-319-04016-5 (eBook)
DOI 10.1007/978-3-319-04016-5
Springer Cham Heidelberg New York Dordrecht London

Library of Congress Control Number: 2014932004

© Springer International Publishing Switzerland 2014

This work is subject to copyright. All rights are reserved by the Publisher, whether the whole or part of the material is concerned, specifically the rights of translation, reprinting, reuse of illustrations, recitation, broadcasting, reproduction on microfilms or in any other physical way, and transmission or information storage and retrieval, electronic adaptation, computer software, or by similar or dissimilar methodology now known or hereafter developed. Exempted from this legal reservation are brief excerpts in connection with reviews or scholarly analysis or material supplied specifically for the purpose of being entered and executed on a computer system, for exclusive use by the purchaser of the work. Duplication of this publication or parts thereof is permitted only under the provisions of the Copyright Law of the Publisher's location, in its current version, and permission for use must always be obtained from Springer. Permissions for use may be obtained through RightsLink at the Copyright Clearance Center. Violations are liable to prosecution under the respective Copyright Law.

The use of general descriptive names, registered names, trademarks, service marks, etc. in this publication does not imply, even in the absence of a specific statement, that such names are exempt from the relevant protective laws and regulations and therefore free for general use.

While the advice and information in this book are believed to be true and accurate at the date of publication, neither the authors nor the editors nor the publisher can accept any legal responsibility for any errors or omissions that may be made. The publisher makes no warranty, express or implied, with respect to the material contained herein.

Printed on acid-free paper

Springer is part of Springer Science+Business Media (www.springer.com)

Preface

> The rabbit runs faster than the fox, because the rabbit is running for his life while the fox is only running for his dinner.
>
> *Aesop*

In many of today's businesses, we are driven by technology and technology changes fast. "Now, here, you see, it takes all the running you can do, to keep in the same place." That is how Alice explains Wonderland in Lewis Carroll's "Through the Looking-Glass". Most engineers think the same about their business environment. Innovation of new technologies is the name of the game. More than USD 1,400 million are spent worldwide to advance technology. We all have to run to keep the strategic positions against our competitors and we have to run to keep the side-effects of our "running", our technological progress, under control.

Over the last 20 years, researching and teaching at the University of St. Gallen and the ETH in Zurich, we have learnt that in order to cope with today's managerial know-how, the technical "Red Queen" Game is a patchwork of successful and failed practices, some theory and a plethora of examples. Learning is not easy since feedback is hampered by good and bad luck. There is a big risk to "rationalize" managerial actions in hindsight.

In this book, we present insights based on personal experience, both in business and academia, grounded as well on over 50 PhD theses on risk management, innovation and logistics. They are not recipes, but rather background information and alternative concepts about how to approach specific questions in technology and innovation management. We hope they are helpful for students in technology management to get a generic understanding on how the world of technology ticks. The experienced manager will hopefully find some stimulus on how to change her or his approach to cope with today's fast moving environment.

In our executive classes where we taught topics from this book, we found that teaching theory, concepts and examples are only part of the game. The far more important learning process and eye-opener is the transfer to real life. Therefore we attached an Economist article with an up-to-date topic to each chapter: The chapter for grasping information, the Economist article for turning that information into applicable thoughts. We hope you will enjoy plenty of these "Aha"-moments that we have also had together with our students.

We would like to thank all the many managers and students for their feedback in the last few years and our many PhDs for their thorough analysis of specific business situations. There were some people that contributed with helpful reviews, graphical illustrations and proof-reading to the book, namely Silvia Boutellier, Annina Coradi, Tamara Aepli, Prof. Dr. Nils Behrmann, Dr. Eric Montagne, and Dr. Nicolas Rohner. Hopefully, this book constitutes some insights for a better understanding of technology and its effects on business, jobs, our environment and last but not least our society.

Zurich, Switzerland   Roman Boutellier
                                                                                                             Mareike Heinzen

Contents

1 Technology Changes Our Life 1
   1.1 Four Different Faces of Technology 2
       1.1.1 The Two-Sided Story 3
       1.1.2 Towards the Positive Effect 5
       1.1.3 Expectations in Technology 6
       1.1.4 Technology Development 7
   1.2 Technology-Driven Business Strategy: Four Basic Questions ... 10
   The Economist: The age of smart machines 11
   References 13

2 Paradigms and Hidden Research Strategies 15
   2.1 Paradigms of Medicine 15
       2.1.1 First Paradigm of Medicine: Hygiene and Vaccines 16
       2.1.2 Second Paradigm of Medicine: Chemicals 17
       2.1.3 Third Paradigm of Medicine: Reduction to
              Information 17
       2.1.4 Fourth Paradigm of Medicine: A Social Construct 18
   2.2 Science Paradigms: Theory of Science 18
   2.3 Methodological Paradigms: Reductionism 20
   2.4 The Linear Model 22
   2.5 Integrated View and Sustainability: A New Paradigm? 23
   The Economist: The race to computerise biology 25
   References 31

3 Technology, Driver of Growth 33
   3.1 Economic Growth 33
   3.2 Technology, the Driver of Growth 36
       3.2.1 Solow's Surprise 38
       3.2.2 Schumpeter: The Entrepreneur 39
       3.2.3 Capitalism 40
   3.3 Technology and Growth in Modern Times 41
   The Economist: The growth machine 42
   References 44

4 Technology Speed and Substitution ... 47
4.1 The Myth of the Increasing Technology Speed ... 47
4.2 The S-Curve ... 51
4.3 Increasing Returns ... 52
4.4 Acceptance of New Technologies ... 53
4.5 Disruptive Innovations ... 54
The Economist: Frugal innovation: Green ink ... 56
References ... 58

5 Diffusion of New Technologies ... 59
5.1 Diffusion of Innovation ... 61
5.2 Standards ... 63
5.3 Industry Structure ... 64
5.4 The Saddle ... 64
5.5 Technology Spreads Like Rumours ... 66
The Economist: Fuel economy ... 67
References ... 70

6 The Myth of the Pioneer ... 71
6.1 Biased Statistics ... 71
6.2 The Early Adopters ... 73
6.3 Benefits from Competitor's Homemade Barriers ... 76
6.4 Switching Costs ... 77
6.5 Entry Barriers ... 77
The Economist: Samsung: The next big bet ... 78
References ... 84

7 Pirates, Pioneers, Innovators and Imitators ... 85
7.1 The Technology Life Cycle ... 85
  7.1.1 Pirates and Pioneers in Ship Technology ... 86
  7.1.2 Pirates and Pioneers in the Music Industry ... 88
7.2 The Lock-in-Effect: The Winner Takes It All ... 89
  7.2.1 Traditional Economics ... 90
  7.2.2 Positive Feedback Loops ... 91
  7.2.3 Effects on the Industry ... 92
The Economist: Virtual currencies: Mining digital gold ... 93
References ... 96

8 Dominant Design Industry ... 97
8.1 Dynamics of Dominant Designs ... 98
  8.1.1 Fluid Phase ... 99
  8.1.2 Transition Phase ... 100
  8.1.3 Distinct Phase ... 100

	8.2	Examples of Dominant Designs	101
		8.2.1 The Video Recorder	101
		8.2.2 Post-It	102
	8.3	Sub-optimal Dominant Designs	103
	The Economist: The lift business: Top floor, please		104
	References		105
9	**Science-Driven Industry**		107
	9.1	Three Types of Research and Development	107
	9.2	Science Versus Technology	109
	9.3	The Law of Big Numbers	114
	9.4	Academic Spin-Offs	115
	The Economist: Global positioning: Accuracy is addictive		116
	References		120
10	**High-Tech Industry**		121
	10.1	From Oil to Fuel Cells?	122
	10.2	The Microprocessor	124
	10.3	Speed of High-Tech	125
	10.4	Subsidies	126
	The Economist: The high-tech industry: Start me up		128
	References		131
11	**Basic Innovation Principles**		133
	11.1	Acceptance of Innovations	133
	11.2	Eight Innovation Principles	135
		11.2.1 No Innovation Outside Strategy	136
		11.2.2 If It Goes Outside the Company's Strategy, Spin It Off	136
		11.2.3 Everyone Knows the Value Proposition	137
		11.2.4 Open One Valve at the Project Start	138
		11.2.5 Go for a Quick Start	139
		11.2.6 Innovation Management = Project Management	139
		11.2.7 Innovation Management = Profit Management	141
		11.2.8 Innovation Management Is People Management	141
	The Economist: Business innovation: Don't laugh at gilded butterflies		142
	Reference		147
12	**Radical Versus Routine Innovation**		149
	12.1	Radical Innovation	150
	12.2	Routine Innovation	151
	12.3	Examples of Radical and Routine Innovations	153
		12.3.1 Float Glass: Dedicated Engineering	153
		12.3.2 Containerization: Global Standards	155

	12.3.3	Teflon: Serendipity at Work	155
	12.3.4	smart: Planning the Unplannable	156
12.4	Drivers of Radical and Routine Innovations		157

The Economist: Free exchange: The humble hero 160
References 162

13 Trend Towards Routine Innovation 163
13.1 Standardization 163
13.2 Miniaturization 165
13.3 Modularity and Co-opetition 167
13.4 Models and Simulation 169
The Economist: Smart systems: Augmented business 169
References 172

14 Design of Industrial Goods 175
14.1 Design Strategies 176
    14.1.1 Inside-Out Design 176
    14.1.2 Outside-In Design 177
    14.1.3 Interaction Design 179
14.2 Design and Art 180
14.3 Design and Product Development 181
14.4 The Designer and Design Costs 183
The Economist: Italian fashion: Dropped stitches 184
References 185

15 Protection of Innovation 187
15.1 Patents 188
    15.1.1 Famous Patent Disputes 189
    15.1.2 From Breadth to Specificity 190
    15.1.3 Patenting Worldwide 190
    15.1.4 Patenting in Europe 192
    15.1.5 Patent Protection 193
    15.1.6 Cross-Licensing 194
15.2 Copyrights and Trademarks 194
15.3 The Protection Mix 196
The Economist: Smart-phone lawsuits: The great patent battle .. 196
References 199

16 Core Capabilities 201
16.1 Managing Core Capabilities 202
    16.1.1 Internal Strengths and Strategic Values 203
    16.1.2 Be Aware of Core Rigidities 203
    16.1.3 Skills and Problem-Solving 205
16.2 Learning 206
The Economist: Technological change: The last Kodak moment? .. 207
References 211

Contents

17 Technical Risk Management 213
    17.1   Construction Failures 213
            17.1.1   The Hyatt Regency, USA 213
            17.1.2   Indoor Swimming Pool in Uster, Switzerland 215
    17.2   Nuclear Disasters 215
    17.3   Engineering Approach to Risk 217
    17.4   Technical Risk Theory 218
            17.4.1   The Hazard or Failure Rate 218
            17.4.2   Reliability 219
            17.4.3   Mean Time Between Failures (MTBF) 220
    17.5   Limitations of the Technical Risk Approach 220
    The Economist: Daily chart: Danger of death! 221
    References 224

18 Human Aspects of Technical Risk Management 225
    18.1   Disputed Technologies 226
    18.2   Risk Perception 226
    18.3   Perrow's Dilemma 229
    18.4   Risk Acceptance 232
    The Economist: Genetic modification: Filling tomorrow's
    rice bowl 233
    References 235

19 Non-proliferation of Adverse Technologies 237
    19.1   Nuclear Power 238
    19.2   Technology and Defence 239
    19.3   NATO's Priorities 240
    19.4   Adverse Technologies 242
    19.5   International Treaties 243
    19.6   Terror and Epidemics 243
    The Economist: Nuclear security: Threat multiplier 244

Glossary 247

Technology Changes Our Life

> We are energized by the great power of technological impact on us. We are intimidated by the magnitude of problems it creates or alerts us to.
> Herbert Simon, Nobel Prize Winner 1978, Economics

Abstract

Technology has more and more impact on our life. It gave us all our wealth, but created and creates many problems as well. We are overwhelmed by its great power and intimidated by the magnitude of the problems it creates.

On September 11, 2001 the world changed (Fig. 1.1): Technology is the key to our wealth, but technology also makes us vulnerable. Thanks to technology we now live much better than all our ancestors, but other people may interrupt our comfortable life with exactly the same technology.

Technology has an influence on our life wherever we go. In developed countries, human beings are confronted daily with dozens of machines and hundreds of various technologies from all over the world: We wake up with the noise made possible by semi-conductors from Korea which are built into plastic shells from Germany. On vacation we take photographs with a camera assembled in Japan, whose lenses come from China and whose batteries are made in Vietnam. The pictures shot are stored with software from India somewhere in the global cloud of Google and sent to our friends with mobile phones made in dozens of different countries. International standards provide worldwide equipment compatibility.

In 2010 administrators in India realized that sacred cows were dying because poor Indians were throwing away the tiny plastic bags used for holding chewing tobacco. Paper bags from Switzerland replaced the plastic bags. The poorest Indian buys products made in rich Switzerland.

At the beginning of the First Gulf War, the American Chief of Staff, Colin Powell, watched his own air force attacking Baghdad in real time on CNN: Reality happens live on television – a technology whose political impact we are only slowly starting to understand. A few years later, the leading country in technology, the US, withdrew its forces from Somalia. American soldiers had clumsy radio transmitting

Fig. 1.1 The day the world changed, September 11th, 2001

apparatus, whereas the Somali combatants were using light mobile phones that communicated via American satellites. The war in Somalia shows how fast technologies become obsolete and how blurred the line between military and civil technology has become. Dual goods, goods used for both military and civilian applications, influence politics. Even world powers have changed or adapted their policies because of the emergence of new technologies. Many old technologies survive thanks to military interventions: The First Gulf War supported the survival of the dominant car engine, the Otto-Engine, a technology that captured its leading position in the 1930s purely by chance. One of the goals of the Second Gulf War was to stabilize a region with the most important raw material, oil. The war was started with the declared mission of destroying dangerous technology – weapons of mass-destruction, as explained by George W. Bush and Tony Blair. More than a decade later US troops were withdrawn from Iraq, oil production is still lower than before the war and no weapons of mass destruction were found. Ten years later fracking-technology makes the US increasingly less dependent on oil imports. The Middle East is losing part of its political importance.

1.1 Four Different Faces of Technology

Arguments about technology's different faces have been with us for thousands of years. In Greek mythology, Daedalus personifies the typical engineer: He uses technology to gain authority. Daedalus was never a major figure in Greek mythology, just an enabler. He is known as the inventor of different tools: saw, axe, plumb line, drill and even glue. In his Metamorphoses, Ovid describes how Daedalus had to engineer solutions to the very specific needs of the king and the queen of Crete: King Minos refused a request from Poseidon to sacrifice his beautiful white bull. In revenge, Poseidon made Minos' wife Pasiphae lust for the bull. She asked Daedalus for help. Daedalus built a wooden cow, put Pasiphae inside and, 9 months later, she gave birth to the Minotaur, the son of Pasiphae and the beautiful bull! The Minotaur had a human body, a bull's head and had to be fed with young Athenians. In order to get rid of this bad "side-effect", Minos asked Daedalus to build a labyrinth in order to imprison the Minotaur. The final solution for the "side-effect" was not technology, but the love between two humans: Theseus, the mythical founder-king of Athens, and Ariadne, the daughter of Minos and Pasiphae. Ariadne helped Theseus to kill her half-brother, the Minotaur. She gave him a sword and a ball of red fleece thread, known as the Ariadne-Thread, which helped Theseus to find the way out of the labyrinth. Ariadne got her idea as well from Daedalus. The Greeks understood already very well the complexity of technology, its side-effects and new problems produced by solutions of old problems.

The second part of the Greek thinking about technology is better known: To prevent his knowledge of the labyrinth from spreading to the public, Daedalus was imprisoned in a tower and was not able to leave Crete by sea. To get out of prison, he invented a flying device (Fig. 1.2), escaped together with his son Icarus who pushed their "air-plane technology" too far. He flew towards the sun, the wax between the feathers of his wings melted. Icarus fell into the sea before his father's very eyes and drowned. Daedalus cried, bitterly lamenting his own engineering arts.

The ancient Greeks' "sex and crime" stories show that discussions about side-effects of old technologies have had the same complexity for thousands of years. They encourage us to consider the long-term consequences of our inventions with great care, lest those inventions do more harm than good. Two thousand and five hundred years ago technology was already perceived as a kind of lust with dangerous side-effects. Losing control may lead to ever more side-effects and finally disasters. A typical modern example is Nuclear Power with its positive effects on climate change and its side-effects made visible in the catastrophe of Fukushima.

1.1.1 The Two-Sided Story

> The next war will be fought with arrow and crossbow.
> *A. Einstein, Nobel Prize Winner in Physics 1921*

Although Marie Curie had discovered radioactivity around 1900, understanding how the reaction took place occurred only years later. However, in 1905 Albert Einstein showed the equivalence of energy and matter in his famous formula $E = m * c^2$. The experiments of Ernest Rutherford, known as the father of nuclear physics, had already shown how much energy is involved in radioactivity. With all that knowledge, the idea was born to use it for energy generation and subsequently the bomb. Finally in 1938, the discovery of nuclear fission of heavy

Fig. 1.2 "Engineer" Daedalus constructing wings for his son Icarus

elements was achieved some months before the outbreak of World War II, following nearly five decades of intensive work on the science of radioactivity and the elaboration of a better understanding of the components of atoms. Uranium 235 decays spontaneously and emits neutrons that activate other uranium nuclei, which again decay. This chain reaction can only be sustained if the neutrons collide with other fissionable material often enough, i.e. the reaction needs a critical mass of Uranium 235.

Seven years after the detection of fission in 1946 a scientist in the US was demonstrating fission techniques to a group of scientists, using a plutonium metal assembly. To control the active mass in the experiment he used a simple screwdriver to reduce the distance between two half-balls of Uranium. He lost control over his screwdriver, the two half spheres came into direct contact and an overcritical mass was achieved much faster than planned. The experiment went out of control. The assembly started to heat up; the scientist was exposed to a dose of 2,100 rem and died 9 days later. Another scientist with the impact of the second highest dose of 360 rem was the only other person involved in the experiment to develop radiation sickness (Messerschmidt, 1990).

Even today the exact mechanism of nuclear fission is not precisely known, although we do know which particles are involved in the reactions and what the half-life of each isotope is. Nevertheless, mankind has succeeded in using chain reactions in wars and in the years following for peaceful purposes as well.

At the beginning of World War II various scientists wrote personal letters to President Roosevelt of the United States. One was signed by Einstein and insistently warned him about the research German scientists were conducting. Roosevelt initiated the Manhattan Project in 1939 that led to the development of the atomic bomb. On August 6, 1945 at 8.15 am the atomic bomb, called "Little boy", created the biggest man-made destruction in human history with 90,000 Japanese killed immediately, 17,000 died in the aftermath and 80 % of the city Hiroshima was destroyed.

Later on, Einstein became a pacifist and for some time was under permanent surveillance by US secret services. Einstein continued to warn people about an uncontrolled use of modern technology and stressed its "side-effects". Today Nuclear Power is still heavily debated, and after the disaster of Fukushima some nations have decided to give up Nuclear Power for good. Some energy-hungry markets like China continue to build new nuclear power stations. They are convinced that the positive aspects far outweigh the negative side-effects.

1.1.2 Towards the Positive Effect

> Therefore we depend on modern technology and new scientific insights – up to the core.
> *M. Eigen, Nobel Prize winner in Chemistry 1967*

Technology can have huge positive effects. Technology brought us today's wealth, long life, health ... in fact, everything. We depend on technology whether we want to accept it or not.

Between 1920 and 1980 Swiss agriculture increased its productivity fivefold due to investments in machines, fertilizers and pesticide. Today farming is highly industrialized and far removed from the rural daily life of the eighteenth century. This high-tech dependence, which is extremely capital intensive, is one of the technical reasons why developed countries have a food surplus and why the developing countries' efficiency in agriculture is very low. Agriculture is a fast-growing industry, a business, which produces machines, fertilizers, pesticides, seeds and genetically modified plants. Since World War II agriculture has expanded and become more specialised, both in the developed and much of the developing world. It is one of the world's largest industries, employing between one and two billion people and producing more than USD 1,500 billion of goods a year.

World output of food per head has gone up by some 25 % over the past 40 years, even though land use has grown by only 10 % and world population has increased by 90 %. Food prices in real terms have fallen by two-fifths, so that consumers in America today, for example, spend on average only 14 % of their household income on food. A success story? Not completely. Modern agriculture is now being shaped by many of the same technologies that are transforming other industries, but is also subject to unique political and economic constraints. It produces an abundance of cheap food in rich countries, but at the same time it heavily pollutes our environment: Meat production in most developed countries has an equally big impact on the atmosphere as public and private traffic together, and over-fertilization has destroyed many drinking water resources.

Agribusiness, a term coined by Ray Goldberg at Harvard Business School in the late 1950s, is a long chain of companies and institutions stretching from the supply of inputs, such as seeds, fertilizer and machinery, to food processors, retailers and end users. In the past, each link of the supply chain did most of its business with its direct neighbours upstream and downstream in the supply chain and had little contact with the rest. Family farmers were central to the system. They decided what to grow and where to sell it, taking the risk and reaping a fair portion of the reward. In 1950, the world's agribusiness was worth USD 420 billion, and farmers added more than a third of the value. By 2028 the market could be worth USD 10 trillion, and some studies assume that farmers' share will fall to 10 % (The Economist, 2000). However, without all these technologies it would not be possible to provide today's population with sufficient food! Therefore every suitable new technology is applied to improve efficiency: The amount of genetically modified food keeps growing. From 1996 to 2005 the global area of genetically modified crops grew to 90 million hectares, 20 times the surface area of Switzerland, despite all the criticism coming from Europe.

Economic research shows that technology is the driving force behind almost all of the increases of productivity and thus growth. This idea is not new. Already Marx was convinced that technology increases productivity, and he was the first who recognized that capitalism is the fertile ground that enables technology to grow. Later on, Schumpeter und Solow refined these ideas and showed that the production factors land, labour and capital explain only a small part of historical increases in production and wealth. Technology is and remains the most important factor for economic growth and progress. Today, technology is accepted by a large majority of people, despite the heavy criticism of seeing progress set equal to technical progress.

1.1.3 Expectations in Technology

> In technology, whatever can be done, will be done.
> *Andrew Stephen "Andy" Grove, former CEO Intel and "the guy who drove the growth phase" of Silicon Valley* (Gaither, 2001)

Once a technology is established, it will be applied and "extrapolated". In university laboratories side-effects can be controlled by well-trained people and in small doses that avoid big catastrophes. Once a technology moves to mass-scale, side-effects do matter and many disasters happen in a climate of trust when users feel too secure. This is why young drivers pay higher insurance premiums than experienced drivers: The younger ones feel over-confident, and thus tend to be involved in more accidents. But experts have the same deficiencies: If one masters the construction of a bridge spanning 30 m, the other tries to build one with a 40 m span. If after a generation of engineering and 20–30 years of technical success, subsequent engineers do not remember the basic considerations of their predecessors anymore, catastrophes are inevitable.

The Tacoma Bridge in Washington, for example, opened to traffic on July 1st in 1940, and dramatically collapsed into the river Puget Sound on November 7th of the same year. At the time of its construction the bridge was the third longest suspension bridge in the world in terms of main span length, following the Golden Gate and the George Washington Bridge. The bridge's collapse had a lasting effect on science and engineering. In today's physics textbooks the event is presented as an example of a catastrophe due to wind resonance. Thus it is known for effects that only become dangerous beyond certain dimensions. Engineering is a human activity with big risks and is therefore heavily regulated. Governments try to foster technical innovation while simultaneously attempting to protect us from its side-effects with rules, laws and standards – a difficult balancing act.

A recent example is Japan, where 600 years ago, people carved stones with Tsunami warnings and moved their buildings to higher regions. Almost 500 years later, modern Japanese trusted new technologies more than traditional precautions. More than 30,000 people were killed in the Tsunami caused by the earthquake in March 2011.

We expect too much from technology: In a US survey 500 people were asked if they would prefer USD 1,000 cash or a free total body computer tomography scan which costs approximately the same today. Seventy three percent of all interviewees preferred a body scan to money. The thinking behind this response is: If there is a diagnosis, there must be a remedy (Schwartz, 2004). We all are full of hope, despite the cases of untreatable cancer and heart diseases we know from our own experience.

The longer the period is between cause and effect, the higher the risks may be. An example is asbestos (Gee and Greenberg, 2002). There was a delay of 50 years between cause, annual asbestos imports to Great Britain, and effect, deadly lung casualties (Fig. 1.3). Therefore many people want to apply the precautionary principle in its strongest form for young technologies, gain experience and only then move to industrial scale applications.

In tiny Switzerland some 20,000 t of asbestos were imported at the peak in the 1980s and some 100 cases per year are reported 40 years later. SUVA, the Swiss insurance for professionals in industry, will pay out some 1.2 billion Swiss francs to cover all cases. These are just the technical figures; the tears and sorrows of all families involved cannot be measured.

1.1.4 Technology Development

Artefacts develop faster than ideas.
*Günther Ropohl, *1939, German philosopher of technology*

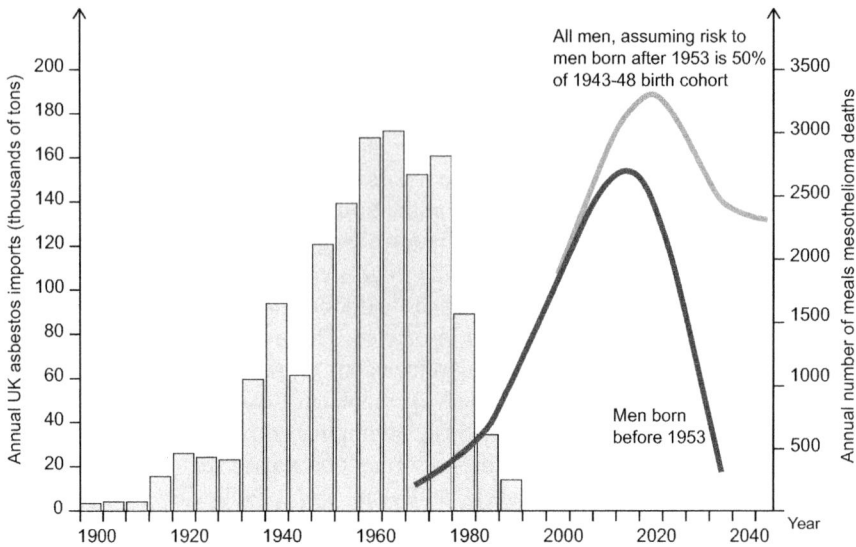

Fig. 1.3 Asbestos in United Kingdom – delay of 50 years between cause and effect (adapted to Gee and Greenberg, (2002))

Everyone who is personally active in technology and tries to solve real problems with whatever is at his or her disposal knows how little is really understood and how quickly the limits of science are reached. Nobody is capable of explaining the extraordinary tone of a Stradivari violin or of giving specific rules on how to construct a violin with the same quality. Or, on the non-technical side, we are only slowly beginning to understand the social effects of television. There is a large risk that technical development outpaces social development. Technology sometimes develops faster than our political system; a big gap opens up between new technologies and safety standards, which is difficult to close. Governments struggle to protect their citizens without undermining innovation.

How technology has changed our lives is evident in every aspect of life. Revolutionary technological developments were achieved through the power of imagination. From the moon landing to a satellite landing on Mars, from the introduction of Microsoft Windows to the introduction of 3G and 4G technologies, from tape recorders to Apple iPods, from few landline telephones to an abundance of Smartphones, SMSs and iPhones, from the origin of the World Wide Web to Web 2.0 and Web 3.0 technologies, from revolutionary internet search engines to social networks, from blogging forums to Internet shopping: Technological developments have impacted our lives in such a way that it is next to impossible to imagine our world without their presence.

However, we have to look at both aspects of development in technology: While technology has simplified our lives with many dreams "just a mouse click away", technology has also led to deteriorating health, psychological problems and stress in a number of people. There are losers, not only winners. Whenever technology changes very fast, new jobs are created but old jobs disappear as well and not all elderly people can be retrained at the speed of the new technology.

In hindsight, most problems of technology are due to their being used the wrong way. Humans, as the inventors of technology, decide if it can and will be used to our peril or to our advantage. It depends on us. But the choice is not always clear: 70 years after Nuclear Power was invented, society is still divided on its value.

Internet technologies have made our world trade and business work faster and easier. It is easier to book flights and railway tickets, and to conduct financial transactions online. The Internet has increased competition in every field. Information and knowledge are accessible everywhere and have become commodities in developed countries. One famous example is Jimmy Wales' Wikipedia platform that offers knowledge for free and is now available in 260 languages.

However, easier and faster also means more work in a shorter amount of time: Smartphones, laptops and the Internet make us available anywhere and anytime, privacy has become a luxury. Emails and chatting have led to much faster communication. Lifestyle habits have changed drastically. Our working life interacts more and more with our free time. The number of hours a person works in her life has increased tenfold from 15,000 h of work 50,000 years before Christ to 150,000 h of work today (Fig. 1.4). We innovate and innovate so that we can work ever more hours!

Is there enough time to learn? Technology itself leads to ever-faster technology development; it is a self-reinforcing process. Whereas it took humans 2 million years to go from stone tools to a Swiss army knife, researchers only needed 50 years to move from the first semiconductor to Fullerene, to move from electronics to nanotechnology. Everybody understands stone tools and army knives. Very few grasp the full impact of electronics, but, undaunted, we move on to the next step: nanotechnology. Many people feel overwhelmed by this fast technological progress and develop a negative attitude towards new technology, like the Amish people in the US.

Even though a technology like the TV had a dramatic impact on democratic participation and access to information in society, the democratic decision process is often slower than technological change. Long before the public is aware of a technology and starts discussing it, scientists have already developed it and are aware of its potential. Political discussions are often a result of public debates: In 1954 DNA was decoded and gene-tech and biotechnology started to boom. Until 1998, when the first sheep, Dolly, was cloned, public awareness of gene technology was very low. After this big bang, public discussion started and politicians are still debating about laws to regulate the new technology. The democratic process needs time, whereas technology moves fast. Nevertheless, many technologies are closely monitored nowadays, as they cause pollution, health problems or could be used in terror attacks. Just a few years ago the EU started a program, "Reach", to make sure that every chemical above 1-t production per year is safe. Side-effects and negative consequences of technological products and processes are under public scrutiny as soon as the press reports some undesired consequences.

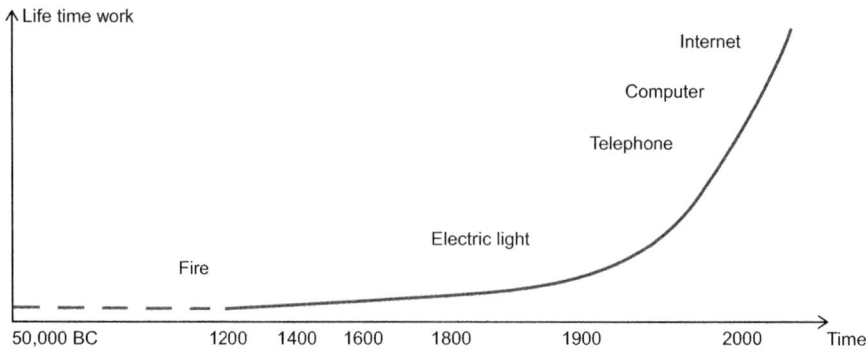

Fig. 1.4 From 15,000 h of work 50,000 years before Christ to 150,000 h of life-time work in 2000 after Christ

While consumers are marvelling at this year's new tablet or Smartphone, researchers are again hard at work developing the next wave of technologies that brings humanity closer to what we know from science-fiction: Precision medicine, self-driving cars, geriatric care, robots intelligence-enhancing tools or 3D-printing in every home. And even while a scientist with some imagination already dismisses these future technologies as old hat or even obsolete, their potential side-effects are still not fully known.

1.2 Technology-Driven Business Strategy: Four Basic Questions

Today, technology mainly exists in enterprises and is developed locally, whereas research is continually shifting to universities and is developed globally. Engineers and managers have to accept that technological progress has to be judged with an integral assessment of several dimensions. Every enterprise should answer four questions with respect to technology (Loveridge, 1990):
1. What will come next?
2. How should management handle new technologies to produce a culture of innovation?
3. What can we do to comply with ethical standards?
4. How can we use technologies to earn money?

The answers can be given only by applying at least three perspectives akin to the simple "Triple Bottom Line": Profit, creation of jobs and environmental effects.
1. Impact on economy
2. Impact on society
3. Impact on the environment

An integral view is needed, a view not taught to most engineers.

Technological questions are of extreme importance today, independent of business or political positions: For strategies, for the efficiency of activities, but also for the survival of mankind.

The Economist: The age of smart machines

Brain work may be going the way of manual work
© The Economist Newspaper Limited, London (May 25th 2013)

IN HIS first novel, "Player Piano" (1952), Kurt Vonnegut foresaw that industry might one day resemble a "stupendous Rube Goldberg machine" (or as Brits would say, a Heath Robinson contraption). His story describes a dystopia in which machines have taken over brainwork as well as manual work, and a giant computer, EPICAC XIV, makes all the decisions. A few managers and engineers are still employed to tend their new masters. But most people live in homesteads where they spend their time doing make-work jobs, watching television and "breeding like rabbits".

It is impossible to read "Player Piano" today without wondering whether Vonnegut's stupendous machine is being assembled before our eyes. Google has designed self-driving cars. America's military-security complex has pioneered self-flying killing machines. Educational entrepreneurs are putting enlightenment online. Are we increasingly living in Vonnegut's dystopia? Or are the techno-enthusiasts right to argue that life is about to get a lot better?

Two things are clear. The first is that smart machines are evolving at breakneck speed. Moore's law—that the computing power available for a given price doubles about every 18 months—continues to apply. This power is leaping from desktops into people's pockets. More than 1.1 billion people own smartphones and tablets. Manufacturers are putting smart sensors into all sorts of products. The second is that intelligent machines have reached a new social frontier: knowledge workers are now in the eye of the storm, much as stocking-weavers were in the days of Ned Ludd, the original Luddite. Bank clerks and travel agents have already been consigned to the dustbin by the thousand; teachers, researchers and writers are next. The question is whether the creation will be worth the destruction.

Two academics at MIT's Sloan Business School, Erik Brynjolfsson and Andrew McAfee, have taken a surprisingly Vonnegutish view on this: surprising because management theorists like to be on the side of the winners and because MIT is one of the great strongholds of techno-Utopianism. In "Race Against the Machine", their 2011 book, they predict that many knowledge workers are in for a hard time. There is a good chance that technology may destroy more jobs than it creates. There is an even greater chance that it will continue to widen inequalities. Technology is creating ever more markets in which innovators, investors and consumers—not workers—get the lion's share of the gains. The Brynjolfsson-McAfee thesis explains one of the most puzzling aspects of the modern economy: why so much technological creativity can co-exist with stagnating wages and mass unemployment.

A new study by the McKinsey Global Institute (MGI), "Disruptive technologies: Advances that will transform life, business and the global economy", shines a light on this problem and produces lots of examples of the way the Internet is revolutionising knowledge work. Law firms are using computers to search through masses of legal briefs and precedents. Financial companies are using computers to monitor news feeds and make financial bets on the basis of the information they uncover. Hospitals are using robots to perform keyhole surgery.

The rate of progress, says MGI, is set to increase dramatically thanks to a combination of Moore's law and the melding of three technologies: machine learning, voice recognition and nanotechnology. Tiny computers will be able to perform jobs once regarded as the peculiar preserve of humans: most middle-class people will soon have access to electronic personal assistants (to book flights or co-ordinate diaries) and wearable physicians (to keep a permanent watch on their vital organs).

MGI puts a typically positive spin on all this. It argues that being spared relatively undemanding tasks will free knowledge workers to deal with more complex ones, making them more productive. It argues that the latest wave of innovation will be good for both entrepreneurs and consumers. Small businesses will be able to act like giant ones, because cloud computing will give them access to huge processing power and storage, and because the Internet destroys distance. Innovators will be able to test their new ideas with prototypes, and then produce them for niche markets. Consumers captured much of the economic gains created by "general-purpose technologies" like steam and electric power, because they stimulated competition as well as increasing efficiency. MGI reckons that so far they have captured two-thirds of the gains from the Internet.

Techno-backlash

Nevertheless, MGI's study has some sympathy with Messrs Brynjolfsson and McAfee. It worries that modern technologies will widen inequality, increase social exclusion and provoke a backlash. It also speculates that public-sector institutions will be too clumsy to prepare people for this brave new world. Policymakers need to think as hard about managing the current wave of disruptive innovation as technologists are thinking about turbocharging it. For one thing, the purpose of education systems, and the skills and knowledge that they impart, will need to be rethought: "chalk and talk" instruction will be done best by machines, freeing teachers to become more like individual coaches to their pupils.

Knowledge-intensive industries will also have to rethink cherished practices. For a start, in an age in which information and processing power are ubiquitous, they will have to become less like guilds, whose reflexes are to regulate supply and restrict competition, and more like mass-market businesses, whose instinct is to maximise the customer base. Innovation will disrupt many areas of skilled work that have so far had it easy. But if we manage them well, smart machines will free us, not enslave us.

References

Gaither, C. (2001, November 12). Andy Grove's tale of his boyhood in wartime. *The New York Times*.

Gee, D., & Greenberg, M. (2002). Asbestos: From "magic" to malevolent "mineral". In P. Harremoës (Ed.), *The precautionary principle in the 20th century*. London, UK: Earthscan. p. 12 ff.

Loveridge, P. (1990). *The strategic management of technological innovation*. Somerset, England: Wiley.

Messerschmidt, O. (1990). Medical aspects of radiation accidents. In C. Streffer, K.-R. Trott, & E. Scherer (Eds.), *Radiation exposure and occupational risks* (pp. 75–95). Berlin, Germany: Springer.

Schwartz, L. (2004). Enthusiasm for cancer screening. *US Journal of American Medical Association, 291*, 71–78.

The Economist. (2000, March 23). A survey of agriculture and technology: Growing pains. *The Economist*, p. 3.

Paradigms and Hidden Research Strategies   2

> ...as we understand the body's repair process at the genetic
> level... we will be able to advance the goal of maintaining
> our bodies in normal function, perhaps perpetually.
> W. Haseltine, CEO Human Genome Science, 2001

Abstract
Many scientists and engineers have a specific world view, a paradigm within which they approach all problems and which helps them to solve their puzzles. From time to time these paradigms change and lead research in new directions. Science and technology seem to be moving in revolutionary steps and long eras of small improvements. Both exert a huge influence on growth!

Albert Einstein is quoted having said that if he had 1 h to save the world: "I would spend fifty-five minutes defining the problem and only five minutes finding the solution". It makes a difference whether you believe that climate change is due to the activity of the sun or whether it is an effect of man's output of CO_2: Problem representations, tacit or explicit, drive our approach in science and technology. These representations may take the form of paradigms, a world view that determines our fundamental approach to problems and their solutions. They depend heavily on industry and are not usually questioned. Medicine is as old as mankind and is therefore a good example to examine and learn from with regard to the nature of the impact it has on such basic paradigms.

2.1 Paradigms of Medicine

For centuries disease was seen as God's punishment. To avoid disease you had to please God or follow the medicine man's rituals in order to lead a decent life. This was a first type of a weak paradigm for diseases. Nevertheless, all societies developed some material remedies by trial and error, mostly based on the use of plants. Nobody understood how these simple drugs worked, nobody knew the

Fig. 2.1 Ignaz Semmelweis: father of the first paradigm of medicine: hygiene

underlying mechanisms: Divergent thinking was common and there was no unifying theory.

2.1.1 First Paradigm of Medicine: Hygiene and Vaccines

In 1860 Ignaz Semmelweis (1818–1865) (Fig. 2.1) had a difficult time convincing medical doctors to wash their hands before treating patients. Infant mortality was high because of poor sanitation in hospitals where women gave birth. Through simple statistics, Semmelweis realized that there was a big difference between the casualties of women and babies in two different hospitals at two different locations. He set up a hypothesis: The difference was due to the simple practice of washing hands. In the location with far more deaths, doctors often visited dead bodies outside the hospital and came back without washing their hands! Semmelweis could not explain his hypothesis; there was no theory and there was no scientific knowledge around to explain his findings. There was only some statistical evidence. Even today, many infections acquired in hospitals correlate with inadequate handwashing.

Only towards the end of the nineteenth century Louis Pasteur (1822–1895) was able to show that many diseases were the effect of some strange, very small agents, bacteria measuring 0.5–5 µm, made visible under the microscope. Even famous Pasteur had great difficulty explaining how such small creatures could kill large organisms such as human beings. But his experiments with wine-fermentation and vaccines against rabies made him such a prominent scientist that his theory of bacteria attacking humans gained increasing acceptance. This led to the first

2.1 Paradigms of Medicine

"scientific" paradigm of disease and defined the principal direction of medical research for many years to come.

Scientists tried and still try to isolate bacteria and find vaccines that kill or prevent disease triggers. Robert Koch (1843–1910) was able to prove that germs trigger tuberculosis and received the Nobel Prize in 1905. Even today tuberculosis still kills more than two million people per year. And we are still looking for better remedies against diseases caused by bacteria.

The theories of Koch and Pasteur led to "convergent" thinking. Universities taught their students what was causing the diseases, and what methods to use in order to obtain better drugs or encourage better behaviour. This led to improved research efficiency. Mainstream research was defined and most researchers tried to get a better understanding of diseases by using standard explanations that were based on Pasteur's theory of bacteria. In Thomas Kuhn's (1977) words: "At least for the scientific community as a whole, work within a well-defined and deeply ingrained tradition seems more productive than work in which no similarly convergent standards are involved." The past success of a paradigm gives researchers the assurance that actual problems may be solved. From time to time new knowledge evolves and a new paradigm may develop.

2.1.2 Second Paradigm of Medicine: Chemicals

Towards the end of the nineteenth century, the chemistry of colours made a big step forward. Coal tar, a waste product of all gas- producing factories, contained many colour pigments. Scientists started to colour objects under the microscope and found that specific colours adhered to specific cells.

For the first time it was possible to watch how a chemical compound, a medicine, moved through a human body. This led to the idea of chemical affinity and later Paul Ehrlich (1854–1905) developed the brilliant idea of "vital-coloration". He coloured cells and tissues from both dead and living animals. Medical doctors were looking for colours that adhered to germs and killed them. The door to the second paradigm was opened: Life is chemistry, a disease is a disturbance of chemical processes in our body, and chemistry can restore the balance. This new theory made it possible to understand why some people had no problems with specific bacteria, and why other people died!

2.1.3 Third Paradigm of Medicine: Reduction to Information

Today, more than 100 years after Pasteur, we think of a disease as an information defect, as a genetic predisposition being triggered by our environment and our lifestyle, an incompatibility between the cell and its environment. Thus a more integrative model drives today's mainstream research in the pharmaceutical industry: Genes and therefore information define diseases. Information not only about our genes, but also information of our lifestyle and our environment are needed in order to come up with an optimal therapy or an effective means of prevention. The newest trend is precision medicine or personalized medicine. Precision diagnostics works in parallel with precision drugs: Depending on a DNA analysis and an assessment of lifestyle, environment and current disposition, an individual therapy per person can be developed. The basic idea that chemistry, genes and information define diseases drives mainstream research in the pharmaceutical industry. The first attempt at personalized medicine is based on the genetic disposition, the third scientific paradigm of medicine. The double mastectomy of Angelina Jolie sent shock waves around the world. She carries the BRCA1 gene, known to be statistically significant for the development of breast cancer.

2.1.4 Fourth Paradigm of Medicine: A Social Construct

The more disease is perceived as being caused by genetic disposition, environment and lifestyle, the more a disease becomes a social construct. Genetics engineering, not only genetics therapy has to be applied as everybody has the right to be restored to health. If beauty belongs to health, everybody has the right to be a good-looking person. Today, a fourth scientific paradigm of medicine is born: Disease as a social construct. Society tells us not only who is sick and who is healthy, but as well who is responsible for it: This opens up a wide and difficult field of discussions about rights or even obligations regarding abortion and death!

2.2 Science Paradigms: Theory of Science

Jürgen Drews, former head of research at Roche in Basel, sees a similarity between pharmaceutical research and Kuhn's (1922–1996) theory of science (Kuhn, 1977): Science and technology move forward in small steps within a paradigm, within a set of assumptions that define our problem representations and thus push our creativity in a specific direction. The paradigm guarantees efficiency: Scientists do not have to question fundamentals all the time. They just solve problems within a proven framework. This leads to hidden research strategies in companies: Engineers develop new products, new solutions for their customers' problems without questioning their generic approaches. When you design a diesel engine and you ask yourself every day whether diesel is the right option or whether you should opt for an electric car, you will lose efficiency.

In Kuhn's early understanding, from time to time this unquestioned progress leads to more and more discrepancies between the predictions and theories generated within a paradigm and the measurements and facts in the real world. Semmelweis' theory about hygiene could not explain why patients who live with a high standard of hygiene get sick and people who live in poor sanitary conditions remain healthy. A diesel strategy may lead someday to smaller and smaller market shares. Once these gaps become too disturbing or even contradictory, some managers or scientists develop new theories and a new paradigm may evolve.

2.2 Science Paradigms: Theory of Science

Science and technology oscillate between mainstream science and revolutionary science. Revolutions change not only the technologies used but may change whole research cultures as well. What is accepted as scientific depends on the reigning scientific community. What is seen as dominant technology depends on the success companies achieve on the market. A new research strategy may be needed.

Of course, not everyone in a field changes because a new paradigm comes on the scene. Some people are so committed to particular paradigms that they will never change. It is not always easy to know when to change and when to abandon those who do not. Matthew, Harrison, & Long (2004) once said: "A new scientific truth does not triumph by convincing its opponents and making them see the light, but rather because its opponents eventually die, and a new generation grows up that is familiar with it".

Newton's theory about the heliocentric model of the cosmos is still taught at school and widely used by engineers. Even the Ptolemaic system is still in use today: If we want to take a photo of next morning's sunrise, we have to make an approximate prediction of the time of this event. The easiest way to do this: The sun turns around the earth, thus the sunrise next morning will be at the same time as today. The prediction will be accurate to a few minutes without any calculation. The use of the Copernican model to solve the same puzzle is far more complicated. However, to solve another problem involving the differences between summer and winter in the northern hemisphere, the Copernican system gives a simple and quick answer.

Historic examples of scientific revolutions cited by Kuhn are the struggle between the Ptolemaic earth-bound system and Copernicus' heliocentric system or the introduction of relativity through Einstein at the beginning of the twentieth century. They have to be observed with caution, as Kuhn later realized.

Kuhn's original model is one of the most often cited philosophical results of the twentieth century. Nevertheless it is being questioned: The system of Copernicus was more complicated than the old Ptolemaic system, since Copernicus stuck to circles! Kepler had to learn three models of our planetary system in parallel when he studied astronomy under the instruction of Michael Mästlin at the University of Tübingen. Kepler's ellipses brought the much sought after unification and simplification. An explanation in the sense of a modern law was given decades later by Newton with his law of gravity. Kepler was deeply disappointed when he learnt that planets move in ellipses and not in perfect circles. He was looking for harmonies in the sky, for signs of an almighty harmonious God. Galileo Galilei supported the Copernican system but rejected Kepler's laws! Copernicus, Kepler and Galileo did not fully understand what they had achieved. Thus Arthur Koestler called them the sleepwalkers (Koestler, 1959), not the first proponents of a new paradigm, but the last great scientists of the old system (Fig. 2.2).

Scientific revolutions need time and old paradigms may survive in parallel with old ones. The same is true with technology: Today we have diesel engines, hybrid and fully electric cars. Many experts try to make predictions which technical paradigm will take over in the next decade.

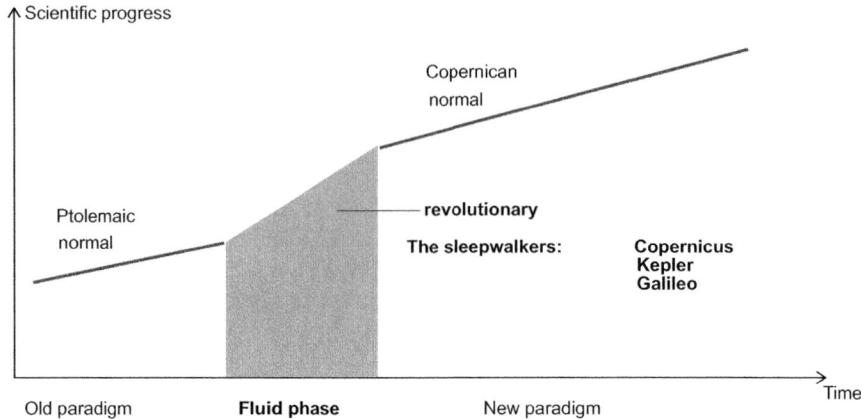

Fig. 2.2 Kuhn's normal and revolutionary science

2.3 Methodological Paradigms: Reductionism

Pharma is an example of paradigm changes and the big success of scientific reductionism. Modern reductionism started with Descartes in 1616. He saw the world as interplay of small particles that can be analysed independently of one another. Once you understand the movement of these small particles, you understand the whole as well because the whole is nothing more than the sum of its parts. Everything, the whole "Res extensa", the "extended world", can be reduced to mechanics: Light emitting from a flower consists of small particles, these hit the eye, set other particles in motion that are perceived by sensors in our eyes and in turn transmitted to the nerves. Even optics is mechanics according to Descartes! He also developed the mathematical tools to describe movement in space, the beginning of analytic geometry. In the few hundred years since Descartes, reductionism and mathematical modelling have become the most powerful tools of modern science (Wilson, 1999). Reductionism is still alive and has even gained momentum through the deciphering of the human genome. Even though reductionism can be very successful – for example simulations shorten times of innovations by factors – mathematical models only answer questions that are explicitly asked! Reductionism becomes very risky when not all important variables are taken into account. Just remember Fukushima or the financial crisis of 2007.

Today we try to explain as much as possible with mathematics. Galileo's view that "The language of nature is mathematics" is widely accepted. This is a first paradigm that drives today's methods in science and engineering: Whenever you have a problem, break it up into smaller parts and introduce abstract mathematical symbols to describe the attributes of these final components given by the observed facts and measured in experiments. This makes it possible to move from one mathematical formula to further ones. They in turn are again better suited for

2.3 Methodological Paradigms: Reductionism

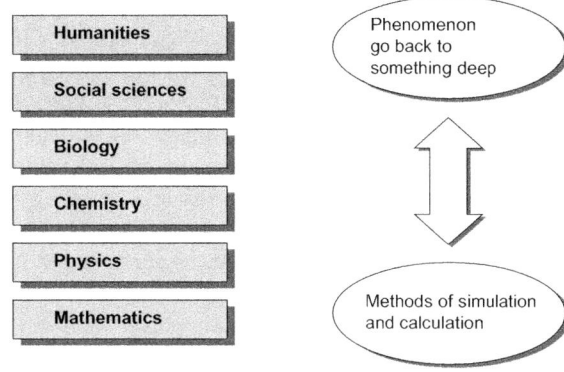

Fig. 2.3 Traditional reductionism: "Hierarchy of science", the two cultures

(Auguste Comte, french rationalist)

observing or applying the mathematical symbols to solve engineering problems. Auguste Comte, a French rationalist, was one of the first scientists to recognize the great advantages of reductionism, the scientific method, and wanted to apply it to social questions as well (Fig. 2.3).

Thomas Kuhn was well aware of this methodological paradigm:

> All of us had been brought up to believe, more or less strictly, in one or another version of a traditional set of beliefs about which I'll briefly and schematically remind you. Science proceeds from facts given by observation. Those facts are objective in the sense that they are interpersonal: they are, it was said, accessible to and indubitable for all normally equipped human observers. They had to be discovered, of course, before they could become data for science, and their discovery often required the invention of elaborate new instruments. But the need to search out the facts of observation was not seen as a threat to their authority once they were found. Their status as the objective starting point, available to all, remained secure. These facts, the older image of science maintained, are prior to the scientific laws and theories for which they provide the foundation, and which are themselves, in turn, the basis for explanations of natural phenomena.
>
> Unlike the facts on which they are based, these laws, theories, and explanations are not simply given. To find them one must interpret the facts – invent laws, theories, and explanations to fit them. And interpretation is a human process, by no means the same for all: different individuals can be expected to interpret the facts differently, to invent characteristically different laws and theories. But, again, observed facts were said to provide a court of final appeal. Two sets of laws and theories do not ordinarily have all the same consequences, and tests designed to see which set of consequences are observed will eliminate at least one of them.
>
> Variously construed, these processes constituted something called the scientific method. Sometimes thought to have been invented in the seventeenth century, this was the method by which scientists discovered true generalizations about and explanations for natural phenomena. Or if not exactly true, at least approximations to the truth. And if not certain approximations, then at least highly probable ones. Something of this sort we had all been taught, and we all knew that attempts to refine that understanding of scientific method and what it produced had encountered deep, though isolated, difficulties that were not, after centuries of effort, responding to treatment. It was those difficulties which drove us to observations of scientific life and to history, and we were considerably disconcerted by that we found there.

Thus Kuhn, a physicist, started to study historical examples and found that most scientists and technicians were solving puzzles within a specific paradigm. By applying the basic principles given in the paradigm, scientists gain more understanding and are able to solve increasingly sophisticated puzzles. This is far more theory-driven than data-driven! Basic science aims for better understanding, applied science solves puzzles dealing with the application of new findings and technology realizes these solutions, thus creating wealth. Kuhn's approach of studying what scientists and engineers actually do showed that our traditional view of science had to be changed. It seems that the pattern of medical paradigms since Pasteur is the pattern we see everywhere: Science and technology move between very efficient mainstream phases and few revolutionary changes that provide new paradigms, new world views. New paradigms neither in science nor in technology are introduced by "proof", but rather by "techniques of persuasion"! Therefore according to Kuhn, and later Feyerabend, lifelong resistance to a new theory may not violate scientific standards. The question is only whether the old theory still provides useful results or not.

Nevertheless, theories share five main characteristics, even in the eyes of Kuhn (1977):

1. Accuracy
2. Consistency
3. Broad scope
4. Simplicity
5. Promise for new research findings.

These criteria may be contradictory. Once again, the choice of a theory is not directly driven by data or hard facts, it is usually a long quarrel and in the end it might be that several theories survive: Newton's theories of mechanics are still used to build cars, even the fastest ones. But it is not the right theory to design a global positioning system! Newton's paradigm and Einstein's paradigm survive in parallel. The importance of the five theory characteristics depends on the problem to be solved. The choice of the "right", not true, theory is a big problem in designing technical systems. Most practitioners are not even aware that they are making a choice. They just know that it works from experience. Engineers do not make decisions based on abstract principles, they produce real systems and check whether they work or not. Their proof of concept is the prototype. Theories and paradigms are just a means to an end and they are used as long as they produce useful results. The paradigms are industry specific, but there is an overarching paradigm still in use, the linear model of Francis Bacon.

2.4 The Linear Model

One of the first manifestations of "modern science" turned out to be the approach of Francis Bacon (1561–1626): Basic science leads to applied science, which leads to technology and with it to economic growth and finally to wealth (Fig. 2.4).

2.5 Integrated View and Sustainability: A New Paradigm?

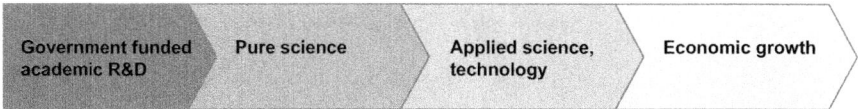

Fig. 2.4 Linear model of Francis Bacon in the sixteenth century: the scientists' optimism

Politicians are still applying this chain of arguments today when they ask for more money for high-tech or when they want to facilitate access to patents for small and medium enterprises (SMEs). The basic goal of modern science, as seen by scientists, engineers and politicians alike, is not knowledge or explanation, but contribution to economic progress. Science has become utilitarian. Society is not pleased with understanding, we want to see wealth. When Faraday told the Chancellor of the Exchequer, William Gladstone, that he had discovered electromagnetic induction, Gladstone asked: "What is the use of your discovery?" Faraday gave the famous answer: "I do not know, Sir, but I'll warrant one day you will tax it."

The linear model is simple and to a certain extent self-explanatory as shown by anecdotal evidence. Nuclear power stations and semi-conductors seem to fit the pattern. Nuclear power was developed based on theoretical arguments at the University of Chicago, where Enrico Fermi (1901–1954) set up the first nuclear reactor. He was able to get a controlled heat-up around Christmas 1942. It took another 9 years until the first reactor near Arco, Idaho/US, produced some electricity, though still at experimental level. The world's first nuclear power plant to generate electricity for a power grid started operations at Obninsk, UDSSR, on June 27th, 1954.

The linear model is certainly not able to explain the complex interplay between society, technology and science. That science can lead to technology is clear, but the reverse is also true: Without technology there would be no modern science: Just think about CERN in Geneva or medical technology such as x-rays and MRI. Science tries to explain existing phenomena; technology and engineering want to create new products or processes, new artefacts.

2.5 Integrated View and Sustainability: A New Paradigm?

Neither reductionism nor sciences create technical products. Many technical objects like churches, bridges or binoculars are technical masterpieces, strongly dependent on aesthetics, such as art. The Sunniberg Bridge near Klosters on the way to Davos (Fig. 2.5), where the economic forum takes place every year, could only have been built by an experienced engineer guided by a strong sense of aesthetics. The engineer behind it, Christian Menn from ETH Zurich, has succeeded in harmonizing technology, material, functionality, environment and users.

The bridge is a masterpiece of simplicity and functionality such as Mario Botta's church in Mogno, Switzerland (Fig. 2.6). Mario Botta uses only local materials and

Fig. 2.5 Sunniberg bridge: realization, not just design

only two of them. This is not calculation, it is a choice driven by human values and aesthetics.

Today technology and science work far more hand in hand than they did 200 years ago: The inventor of the locomotive, Stephenson, was an illiterate and had no idea about contemporary science. His machines even contradicted some laws of thermodynamics of that time, but today's chip technology needs scientific know-how in order to develop newer generations of chips. Until the mid-nineteenth century, there was a big gap between science and technology, the application of science.

A very good example of this gap are our patent laws: When the modern patent regulations were created in the US and Europe some 200 years ago, lawyers introduced the rule that for every patent to be approved, a potential application of the technical idea had to be included in order to make sure that scientific results would not be patented. Science should be open to everybody and science was defined in those times as know-how without technical application, without any direct use.

This has changed completely! Today's science is in many fields so close to application that most universities promote start-ups and businesses based on their founders' own research results. In life sciences and software there seems to be no big time delay anymore between scientific inventions and business applications.

Increasingly, science is being seen as the driver behind technology and, indirectly, technical progress which is equivalent to general progress in the view of many people. There is some opposition, however, the consequences of new

Fig. 2.6 Mario Botta's church in Mogno, Switzerland

technology have been too devastating to be overlooked. Throughout history people tried to live a decent life. Progress is a modern idea and is defined by society, not engineers. We are looking more and more for products that integrate and balance natural sciences, social sciences, environment and aesthetics. But sometimes old ideas are coming back and technical optimism based on breakthrough scientific results take over.

The Economist: The race to computerise biology

Bioinformatics: In life-sciences establishments around the world, the laboratory rat is giving way to the computer mouse—as computing joins forces with biology to create a bioinformatics market that is expected to be worth nearly $40 billion within three years

© The Economist Newspaper Limited, London (Dec 12th 2002)

FOR centuries, biology has been an empirical field that featured mostly specimens and Petri dishes. Over the past five years, however, computers have changed the discipline—as they have harnessed the data on genetics for the pursuit of cures for disease. Wet lab processes that took weeks to complete are giving way to digital research done *in silico*. Notebooks with jotted comments, measurements and drawings have yielded to terabyte storehouses of genetic and chemical data. And empirical estimates are being replaced by mathematical exactness.

Welcome to the world of bioinformatics—a branch of computing concerned with the acquisition, storage and analysis of biological data. Once an obscure part of computer science, bioinformatics has become a linchpin of biotechnology's progress. In the struggle for speed and agility, bioinformatics offers unparalleled efficiency through mathematical modelling. In the quest for new drugs, it promises new ways to look at biology through data mining. And it is the only practical way of making sense of the ensuing deluge of data.

The changes wrought by computers in biology resemble those in the aircraft and car industries a decade or so ago, after the arrival of powerful software for CAD (computer-aided design) and CFD (computational fluid dynamics). In both industries, engineers embraced the new computational modelling tools as a way of designing products faster, more cheaply and more accurately. In a similar way, biotech firms are now looking to computer modelling, data mining and high-throughput screening to help them discover drugs more efficiently.

In the process, biology—and, more specifically, biopharmacy—has become one of the biggest consumers of computing power, demanding petaflops (thousands of trillions of floating-point operations per second) of supercomputing power, and terabytes (trillions of bytes) of storage. Bioinformatics is actually a spectrum of technologies, covering such things as computer architecture (eg, workstations, servers, supercomputers and the like), storage and data-management systems, knowledge management and collaboration tools, and the life-science equipment needed to handle biological samples. In 2001, sales of such systems amounted to more than $12 billion worldwide, says International Data Corporation, a research firm in Framingham, Massachusetts. By 2006, the bioinformatics market is expected to be worth $38 billion.

The opportunity has not been lost on information technology (IT) companies hurt by the dotcom bust and telecoms meltdown. Starting in 2000, IBM was the first to launch a dedicated life-sciences division. Since then, a host of other IT firms have jumped aboard the bioinformatics bandwagon. Along with IBM, Sun Microsystems has staked a claim on the computing and management part of the business. Firms such as EMC and Hewlett-Packard have focused on data storage. Agilent, SAP and Siebel provide so-called decision support. Even build-to-order PC makers such as Dell have entered the fray with clusters of cheap machines.

A swarm of small start-up firms has also been drawn in, mostly to supply data, software or services to analyse the new wealth of genetic information. Companies such as Accelrys in San Diego, California, Spotfire in Somerville, Massachusetts, and Xmine in Brisbane, California, are selling software and systems to mine and find hidden relationships buried in data banks. Others such as Open Text of Waterloo, Ontario, and Ipedo in Redwood City, California, have built software that improves communication and knowledge management among different areas of pharmaceutical research. Gene Logic of Gaithersburg, Maryland, has created a business to collect samples and screen their genetic code for proprietary research libraries.

And Physiome Sciences of Princeton, New Jersey, is providing computer-based modelling systems that offer an insight into drug targets and disease mechanisms.

Bioinformatics is not for the faint of heart, however. Over the past year, the fortunes of a number of biotechnology firms have faltered, as venture-capital funds have sought alternative investments. Venerable names of biotechnology, including Celera Genomics of Rockville, Maryland, LION Bioscience of Heidelberg, Germany, and others, have found themselves scrambling to change the way they do business. Yet, for all the turbulence in the industry, the bioinformatics juggernaut remains on track, fuelled by new forces changing the pharmaceutical industry.

Gene genie

In retrospect, the marriage of genetics and computers was pre-ordained. After all, biotechnology is based on the genetic building-blocks of life—in short, on nature's huge encyclopedia of information. And hidden in the vast sequences of A (adenine), G (guanosine), C (cytosine) and T (thymine) that spell out the genetic messages—ie, genes—are functions that take an input and yield an output, much as computer programs do. Yet the computerisation of genetics on such a grand scale would not have occurred without the confluence of three things: the invention of DNA microarrays and high-throughput screening; the sequencing of the human genome; and a dramatic increase in computing power.

"In just a few years, gene chips have gone from experimental novelties to tools of the trade."

More commonly known as "gene chips", microarrays are to the genetic revolution of today what microprocessors were to the computer revolution a quarter of a century ago. They turn the once arduous task of screening genetic information into an automatic routine that exploits the tendency for the molecule that carries the template for making the protein, messenger-ribonucleic acid (m-RNA), to bind to the DNA that produces it. Gene chips contain thousands of probes, each imbued with a different nucleic acid from known (and unknown) genes to bind with m-RNA. The resulting bonds fluoresce under different colours of laser light, showing which genes are present. Microarrays measure the incidence of genes (leading to the gene "sequence") and their abundance (the "expression").

In just a few years, gene chips have gone from experimental novelties to tools of the trade. A single GeneChip from Affymetrix, the leading maker of microarrays based in Santa Clara, California, now has more than 500,000 interrogation points. (For his invention of the gene chip, Affymetrix's Stephen Foder won one of *The Economist*'s Innovation Awards for 2002.) With each successive generation, the number of probes on a gene chip has multiplied as fast as transistors have multiplied on silicon chips. And with each new generation has come added capabilities.

The sequencing of the human genome in late 2000 gave biotechnology the biggest boost in its 30-year history. But although the genome sequence has allowed more intelligent questions to be asked, it has also made biologists painfully aware of how many remain to be answered. The genome project has made biologists appreciate the importance of "single nucleotide polymorphism" (SNP)—minor variations in DNA that define differences among people, predispose a person to disease, and influence a patient's response to a drug. And, with the genetic make-up of humans broadly known, it is now possible (at least in theory) to build microarrays that can target individual SNP variations, as well as making deeper comparisons across the genome—all in the hope of finding the obscure roots of many diseases.

The sequencing has also paved the way for the new and more complex field of proteomics, which aims to understand how long chains of protein molecules fold themselves up into three-dimensional structures. Tracing the few thousandths of a second during which the folding takes place is the biggest technical challenge the computer industry has ever faced—and the ultimate goal of the largest and most powerful computer ever imagined, IBM's petaflop Blue Gene. The prize may be knowledge of how to fashion molecular keys capable of picking the lock of disease-causing proteins.

The third ingredient—the dramatic rise in computing power—stems from the way that the latest Pentium and PowerPC microprocessors pack the punch of a supercomputer of little more than a decade ago. Thanks to Moore's law (which predicted, with remarkable consistency over the past three decades, that the processing power of microchips will double every 18 months), engineers and scientists now have access to unprecedented computing power on the cheap. With that has come the advent of "grid computing", in which swarms of lowly PCs, idling between tasks, band together to form a number-crunching mesh equivalent to a powerful supercomputer but at a fraction of the price. Meanwhile, the cost of storing data has continued to fall, and managing it has become easier thanks to high-speed networking and smarter forms of storage.

Banking on failure

Despite such advances, it is the changing fortunes of the drug industry that are pushing biology and computing together. According to the Boston Consulting Group, the average drug now costs $880m to develop and takes almost 15 years to reach the market. With the pipelines of new drugs under development running dry, and patents of many blockbuster drugs expiring, the best hope that drug firms have is to improve the way they discover and develop new products.

Paradoxically, the biggest gains are to be made from failures. Three-quarters of the cost of developing a successful drug goes to paying for all the failed hypotheses and blind alleys pursued along the way. If drug makers can kill an unpromising approach sooner, they can significantly improve their returns. Simple mathematics shows that reducing the number of failures by 5% cuts the cost of discovery by nearly a fifth. By enabling researchers to find out sooner that their hoped-for compound is not working out, bioinformatics can steer them towards more promising candidates. Boston Consulting believes bioinformatics can cut $150m from the cost of developing a new drug and a year off the time taken to bring it to market.

That has made drug companies sit up. Throughout the 1990s, they tended to use bioinformatics to create and cull genetic data. More recently, they have started using it to make sense of it all. Researchers now find themselves swamped with data. Each time it does an experimental run, the average microarray spits out some 50 megabytes of data—all of which has to be stored, managed and made available to researchers. Today, firms such as Millennium Pharmaceuticals of Cambridge,

Massachusetts, screen hundreds of thousands of compounds each week, producing terabytes of data annually.

The data themselves pose a number of tricky problems. For one thing, most bioinformatics files are "flat", meaning they are largely text-based and intended for browsing by eye. Meanwhile, sets of data from different bioinformatics sources are often in different formats, making it harder to integrate and mine them than in other industries, such as engineering or finance, where formal standards for exchanging data exist.

More troubling still, a growing proportion of the data is proving inaccurate or even false. A drug firm culls genomic and chemical data from countless sources, both inside and outside the company. It may have significant control over the data produced in its own laboratories, but little over data garnered from university research and other sources. Like any other piece of experimental equipment, the microarrays themselves have varying degrees of accuracy built into them. "What people are finding is that the tools are getting better but the data itself is no good," says Peter Loupos of Aventis, a French drug firm based in Strasbourg.

To help solve this problem, drug firms, computer makers and research organisations have organised a standards body called the Interoperable Informatics Infrastructure Consortium. Their Life Science Identifier, released in mid-2002, defines a simple convention for identifying and accessing biological data stored in multiple formats. Meanwhile, the Distributed Annotation System, a standard for describing genome annotation across sources, is gaining popularity. This is making it easier to compare different groups' genome data.

Tools for the job

Such standards will be a big help. One of the most effective tools for probing information for answers is one of the most mundane: data integration. Hence the effort by such firms as IBM, Hewlett-Packard and Accelerys to develop ways of pulling data together from different microarrays and computing platforms, and getting them all to talk fluently to one another. A further impetus for data integration, at least in America, is the Patent and Trademark Office's requirement for filings to be made electronically from 2003 onwards. The Food and Drug Administration is also expected to move to electronic filing for approval of new drugs.

It is in data mining, however, where bioinformatics hopes for its biggest pay-off. First applied in banking, data mining uses a variety of algorithms to sift through storehouses of data in search of "noisy" patterns and relationships among the different silos of information. The promise for bioinformatics is that public genome data, mixed with proprietary sequence data, clinical data from previous drug efforts and other stores of information, could unearth clues about possible candidates for future drugs.

Unlike banking, bioinformatics offers big challenges for data mining because of the greater complexity of the information and processes. This is where modelling and visualisation techniques should come in, to simulate the operations of various biological functions and to predict the effect of stimuli on a cell or organ. Computer modelling allows researchers to test hunches fast, and offers a starting-point for further research using other methods such as X-ray crystallography or spectroscopy. It also means that negative responses come sooner, reducing the time wasted on unworkable target drugs.

Computational models have already yielded several new compounds. BioNumerik of San Antonio, Texas, has modelled the way certain drugs function within the human body. It has also simulated a specific region of a cell to observe the interaction between proteins and DNA. Thanks to its two Cray supercomputers running simulations that combine quantum physics, chemistry and biological models, BioNumerik has been able to get three compounds into clinical trials. Frederick Hausheer, BioNumerik's founder and chief executive, expects his firm's modelling technology to cut the time for discovering new drugs by a third to a half.

In a similar vein, Aventis has used several models of cells and disease mechanisms to discover a number of new compounds. And Physiome Sciences now sells a product that integrates various clinical and genomic data to generate computer-based models of organs, cells and other structures.

"A big risk of computer modelling and other tools is to rely too much on them."

For all their power, these computer modelling techniques should be taken with at least a grain or two of salt. Although they allow researchers to tinker with various compounds, they will never replace clinical trials or other traditional methods of drug research. Even monumental bioinformatics efforts, such as the Physiome Project, will only help researchers refine their ideas before getting their hands wet. "If people haven't done this kind of work before, they won't understand how difficult it really is," says Dr Hausheer.

Indeed, a big risk of computer modelling and other information tools is to rely too much on them, says Martin Gerstel, chairman of Compugen, a maker of analytical and interpretation tools based in Jamesburg, New Jersey. Many researchers confuse the data generated by bioinformatics with information. The danger with all the computing power being brought to bear is that it is becoming seductively easy for biologists to rely on the number-crunching potential of computers and to ignore the scientific grind of hypothesis and proof. As the technology of bioinformatics outpaces the science of biology, the science risks becoming a "black box", the inner workings of which few will be able to comprehend.

To avoid this, biologists need an ever broader set of skills. For instance, the most pervasive impact of information technology on biology has been through wholesale quantification. Suddenly, biologists are being forced to become mathematicians in order to describe the biological processes and models involved. That implies a demand for wholly new sets of skills and educational backgrounds.

Such changes are not unlike those that affected physics and chemistry during the 1940s, when new computational paradigms created the foundations for nuclear energy and the age of plastics. Much as computing made physics more predictable and prolific, many believe that its new alliance with mathematics will make biology more powerful and definitive. But the marriage will experience some turbulent times before achieving the full flowering of its promise.

References

Koestler, A. (1959). *The sleepwalkers*. New York, NY: Macmillan.
Kuhn, T. (1977). *The essential tension*. Chicago, IL: The University of Chicago Press.
Matthew, H.C.E., Harrison, B., & Long, R.J. (2004). *The Oxford dictionary of national biography*. Oxford: Oxford University Press (p. 596).
Wilson, E. (1999). *Consilience: The unity of knowledge*. New York, NY: Vintage Books.

Technology, Driver of Growth

3

> *The proposition that growth in itself creates value is so deeply entrenched in the rhetoric of business that it has become an article of almost unquestioned faith that growth is a good thing.*
> Richard Rumelt, author of Good Strategy, Bad Strategy

Abstract

Growth has been the single most important characteristic of global economy for the last 200 years. It is very much driven by technology. Recent estimates set its impact at 35–55 % of total growth. But technology needs a specific social environment in which to thrive: Capitalism produced entrepreneurs and property rights, two basic pillars of innovation and growth.

3.1 Economic Growth

Since 1900, real GDP has grown by a factor of 30 in the US. Paul Samuelson, Nobel Prize Winner, thinks: "This is perhaps the central economic fact of the century." (Samuelson & Nordhaus, 2005). Continued growth has enabled advanced industrial economies to provide more of everything to their citizens. In the long run, economic growth is the single most important factor in the economic success of a nation. The growth rate of output per person determines the rate at which a country's economic standard of living rises. With 1 % of GDP growth, 2 % of poverty can be reduced (Fig. 3.1).

Productivity, doing more with fewer resources, remains at the centre of all economic activities. Business growth is similar. The value of individual companies depends heavily on two parameters: Free cash flow and future growth. Some precaution is needed in business growth and growth of economies. It should be sustainable. Over the last century the GDP of major high-income countries such as the United States, Germany, France and Japan has grown sharply. Output has increased faster than labour input, reflecting upturns in capital and technological advances. Between 1870 and 2000 the GDP per capita of the 16 highest income countries grew on average 2.8 % per year, the GDP per work hour 2.3 %, total hours worked in a lifetime 0.5 % and labour force 1.0 % (Samuelson & Nordhaus, 2005).

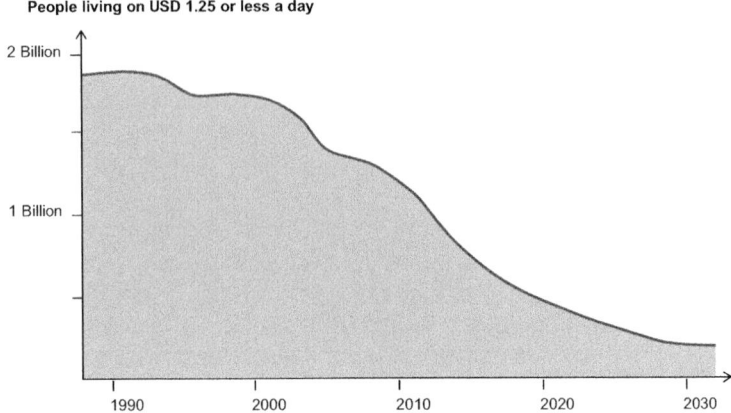

Fig. 3.1 One percent of GDP growth reduces poverty by 2 % (The Economist, 2013)

GDP per work hour is an important metric as it reflects the change in productivity and thus has a close correlation with an increase in living standards. But what are the major sources behind this growth?

Many successful countries have had different growth paths: While Britain became the world economic leader in 1800 by pioneering the Industrial Revolution, Japan joined the growth race later by first successfully imitating technologies and then building its own expertise in car manufacturing and electronics. The same story is being repeated today in China, India, Russia and Brazil, the BRIC countries. They all profit from existing technologies and managerial know-how and have growth rates far above the countries they are copying. At any rate, it seems to be very difficult to achieve high growth rates if you are one of the very wealthy countries: Copying is no option when an economy is in a leading position. Trial and error cannot be avoided, but is much less efficient. Therefore growth rates decline.

Economies change their structure over the years. In industrialized economies the service sector has been growing. The current list of Fortune 500 companies contains more service companies and fewer manufacturers than in previous decades. Within the past 300 years the agricultural sector has lost 60 % of its share of GDP, whereas the service sector has increased its share to 50 %. The service economy in developed countries consists mainly of financial services, health, and education. A "servitization" of products has taken place. Today, many products are being or have been transformed into services. For example, IBM treats its business as a service business, even though it still manufactures computers. In leading economies the change in economic structures creates approximately 2 % new jobs per year and makes 2 % of older jobs redundant. When talking about growth, 2 % seems to be a frequently occurring number. The worldwide population also has a steady yearly growth of approximately 2 %. Many mature industries achieve some 2 % productivity increase per year. Leading economies grow about 2 % per year.

However, it is not only "servitization" that leads to a change in economic growth. The demography of a country is also an important factor. Growth triggered by rapid technological adoption is often associated with a young population as we have in Southeast Asia and the US or with a society that is reshaping after a war which destroyed the economic power of established leading companies. Examples are Europe and Japan after World War II. The fact that old technologies persist in the face of new developments can also be explained by demography: Elderly people have more power in industry and society compared with the younger and less experienced population. For this reason it is easier for a country without existing industrial power to adopt new technologies than one with a successful and thus persistent domestic industry. This fact helped China to emerge over the last four decades. It will be interesting to see how China copes with its youngest generation in the future. They have to support two generations, parents and grandparents. Growth is expected to slow down.

However, we have to be careful when drawing the conclusion that younger societies will have a higher economic growth. For example, Bils and Klenow (2000) show mixed results: Schooling and the education of young people does not always lead to growth. Nevertheless, the reverse was empirically validated: The more the economy of a country grows, the better educated its inhabitants are. One explanation of this observation is that the knowledge and experience of older workers is still needed in order to guide the creativity and ideas of young people.

Even though the growth paths of different nations are very diverse, particularly for developing countries like China and India, four major factors play an important role:

1. Natural resources, such as raw materials, land, minerals, fuels, environmental quality
2. Human resources, such as labour supply and quality of workers and their degree of discipline
3. Capital formation, such as factories and infrastructure
4. Level of technology, such as knowledge and managerial know-how

The relationship between all four sources is often described in the *aggregate production function* (APF), where Q is the output, C is productive capital, L is labour inputs, and R stands for natural resource inputs. A represents the level of technology in the economy and F represents the production function. Output is a function of capital, labour and natural resources multiplied by the factor technology (Fig. 3.2).

With this formula, technology has the role of increasing the productivity of the input. Productivity is the ratio of output to a weighted average of inputs. Technology improves through new inventions or the adoption of technologies from abroad and allows a country to produce more output with the same level of inputs (Samuelson & Nordhaus, 2005).

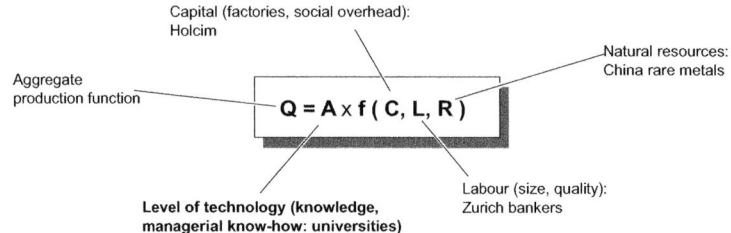

Fig. 3.2 GDP growth has four major sources: technology seems to be the most important

3.2 Technology, the Driver of Growth

Technology is certainly one of the most important drivers in today's economy. Remember the flood of innovations we have had over the last 130 years! Technological change happens through process innovations or the introduction of new products or services. Examples of process innovations that have greatly increased productivity were the steam engine, the generation of electricity, the photocopy machine and the Internet. Fundamental product innovations include telephones, radios and televisions and computers. For the most part, these innovations developed in a continuous stream of small improvements. Because telecommunications, computers and the Internet have dramatically contributed to our rise in living standards, scientists, economists, engineers and politicians alike believe in the positive effects of future innovations and are always looking for ways to encourage technological progress. They have found that technological change is not only a mechanical process of simply finding better innovations, but is driven by entrepreneurial spirit. We often forget that tinkering is one of the most important drivers of inventions, while public acceptance is the most important factor of selection.

Science needed some 200 years after Galileo Galilei to become relevant for technological innovation: Between 1750 and 1800 technology started to be science-driven instead of trial and error. This resulted in a sharp increase in GDP per person in Western Europe from 1850 onwards (Fig. 3.3).

Technological change driven by scientists or inventors has been treated for many years as something that mysteriously appears in economics. Recent research has begun to focus on the sources of technological change. This research has sometimes been called "new growth theory" and it tries to explain how endogenous forces such as investments in human capital, knowledge and innovation can influence technological change. This theory also emphasizes that technology is an output of the economic system that can be subject to market failures such as positive externalities or heavy state-subsidies and spill-over effects because technology is a public good that is expensive to produce but cheap to reproduce.

Some important drivers of technology today are:
- *Technology:* New technology evolves from old technologies through continuous improvement. Typical examples are diesel engines for cars.

3.2 Technology, the Driver of Growth

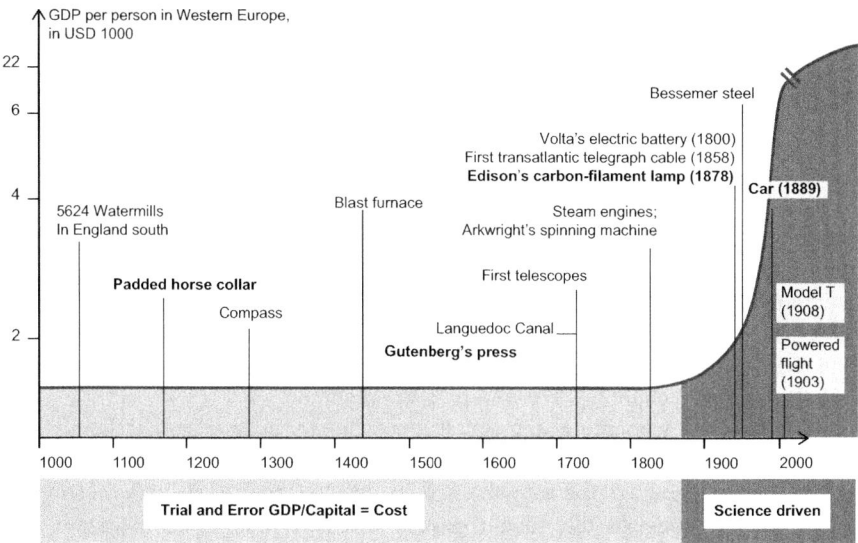

Fig. 3.3 GDP growth: driven by innovation, technology after mid-nineteenth century

- *Science:* It still takes a long time for a product to move from the first results in the lab to the main market. Politicians perceive this as a handicap, try to change it and therefore subsidize science.
- *Problems:* We are forced to solve the problems of today's technology: Toxic gases from cars, nuclear waste, global warming or food shortages. Most people expect the solution to come from even more technology. Technology drives technology.
- *Fashion:* Fashion drove SMS, iPods, and even gothic cathedrals. Designs often depend on sophisticated technologies.
- *Cost reduction*: In a global market with stiff competition it is essential to lower costs every day. Car OEMs probably have the most experience anywhere in driving down costs.
- *Laws:* The public tries to push technology in specific directions with laws: Sometimes through standards, sometimes through bans or subsidies. Asbestos had to be replaced by concrete. Photovoltaics and lean technology are heavily subsidized.

Miniaturization has made possible the construction of new machinery in much shorter time than in the past: You buy the components on the market and create a new product, a new system without using a lot of resources. This has led to "the maker-movement" of the early twenty-first century: Many people develop new machinery or new application software in their spare time. This movement is supported by a vague sense of duty to "save the world", be it from terrorism or global warming. To a certain degree, these basic trends explain the increase in the number of engineering students at many technical universities over the last 10 years. Engineering has become far more attractive when you are able to produce a complete system within a master's thesis! Robots are far more interesting than cogs!

3.2.1 Solow's Surprise

Robert M. Solow, an American economist born in 1924, was surprised that technology made such a difference in growth. In his model, Solow separated the determinants of growth into capital, labour and technical progress. With his model, he empirically showed that sustainable economic growth could only be explained by technical progress. It is not the capital driving the machine that drives growth, but it is technology that improves the machine.

With a growth accounting approach it is possible to separate capital, labour, and technical progress and determine the individual contributions of each factor. The fact that three quarters of the national income goes to labour while only one quarter goes to capital suggests that labour growth will contribute more to output than capital growth. Solow was able to empirically prove this relationship. Output growth per year follows the fundamental equation of growth accounting:

$$Output\ growth = \frac{3}{4} labour\ growth + \frac{1}{4} capital\ growth + technology\ growth$$

This is Solow's surprise: Technology makes the difference, not capital! This formula is based on historical data of the US economy and differs from one economy to the next.

As the impact of technology cannot be measured directly, it is a residual parameter. It explains the remaining difference when everything else has been measured.

From 1900 to 2003 the average output in growth in the US was 3.4 %. By subtracting capital and labour with their weights given in Solow's equation, the contribution of technology to growth is 1.7 %. This can be seen as an economical definition of technology. It is much broader than what engineers understand when they talk about technology. It is everything that contributes to growth and is neither capital nor labour!

People at the beginning of the twentieth century could not co-operate efficiently over longer distances; they had a logistics problem that could not be solved by technology only. Starvation was, and still is, a logistics problem, not a problem of raw materials. Transportation infrastructure had to be developed. Thus political, legal and economic innovations were needed in order to deploy the full potential of those technologies. August von Hayek believed that the protection of private property, the acceptance of contracts and the extension of the notion of "we" were at least as important as any technical innovation to develop global trade. William Easterly, an American economist, has observed growth in developing countries and comes to very similar conclusions:

Economic growth occurs when people have the incentive to adopt new technologies, being willing to sacrifice current consumption while they are installing the new technology for future payoff. This leads to a steady rise over time in the economy's productive potential and people's average income. The incentives that are important are the same as I've already discussed. Good government that doesn't steal the fruit of workers' labour is the essence of it. The Romans and the Chinese had centralized authoritarian governments that devoted most of their resources to war and bureaucracy. The Roman Empire thought of production as something to be left to the slaves, not a good attitude for technological progress. Nineteenth- and twentieth-century America had (and has) a vibrant private market that rewarded the inventors of new and improved lighting. Ecuador, Costa Rica, Peru, and Syria all had unpredictable government policies that tended to discourage investment in the future through innovation. So we reach the same old conclusion: incentives matter for growth. But there are a few complications about incentives for technological progress. Technological progress creates winners and losers. Beyond the happy facade of technological creation are some technologies and goods that are being destroyed. Economic growth is not simply more of the same, producing larger and larger quantities of the same old goods. It is more often a process of replacing old goods with new goods. People who were producing the old goods may well lose their jobs, even as new jobs – probably for other people who lost their jobs – are created producing the new goods. In the United States, for example, around 5% of jobs are destroyed every three months, with a similar number of new jobs created. Vested interests wedded to the old technology may want to block new technologies.

(Easterly & Easterly, 2001)

Growth depends after all on society, on human beings and not on technology alone. Some human beings are especially important for growth: They are called entrepreneurs.

3.2.2 Schumpeter: The Entrepreneur

Schumpeter sees the entrepreneur as the real driver behind growth (Schumpeter, 1942). Thus some researchers talk about Schumpeterian growth. The entrepreneur shifts resources out of fields with a low yield into fields with a higher yield. Entrepreneurial spirit changes market structures, which leads to earnings and losses or jobs that are created and destroyed. Earnings are used to secure research and development (R&D) of big corporations and universities. This R&D, in turn, leads to new innovations, a self-perpetuating cycle, with Kuhnian revolutions that are called storms by Schumpeter (Fig. 3.4).

Schumpeter's everlasting storms of innovations disrupt market structures, lead to new forms of production and finally to growth. Schumpeter sought to show that innovation-originated market power could provide better results than socialism. He argued that technological innovation often created temporary monopolies, allowing higher profits that would soon be 'stolen' by rivals and imitators. These temporary monopolies were necessary in order to provide the incentives needed for companies to develop new products and processes, to pay for big R&D.

Schumpeter was convinced that monopolies are good for innovation and that big companies invest more into R&D than small and medium enterprises (SMEs). This view is in doubt today. Yes, there are industries that need big companies in order to

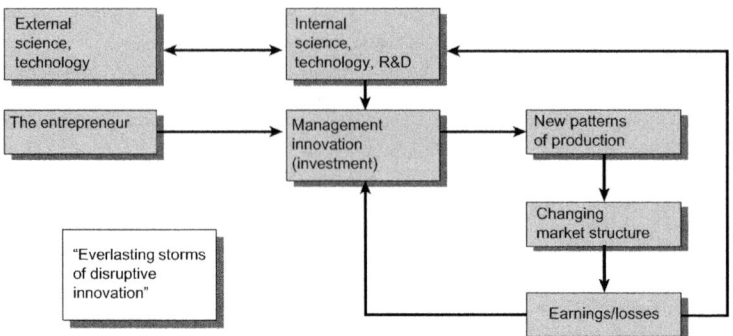

Fig. 3.4 Schumpeter (1942): innovation drives growth and entrepreneurs drive innovations

grow: Wherever you have unpredictability combined with high risks, you need big companies who are able to survive unavoidable failures. The pharmaceutical industry is a good example. SMEs are well suited to developing new potential drugs, but big companies are needed to carry out the very expensive tests done, to obtain the approvals of the authorities, and to distribute the drugs.

3.2.3 Capitalism

The social system that has turned out to be very efficient in promoting innovation and technology is capitalism. In a capitalistic society, innovation means life or death for every firm because time is money. Thus innovators are heavily promoted, are well paid and their inventions are well protected through property rights. Karl Marx was convinced that humanity had to go through a phase of capitalism in order to create the necessary wealth before being able to introduce socialism. The US economist Baumol (2002) called capitalism the "machine whose primary product is economic growth".

The growth performance of capitalist economies is mostly attributed to competitive pressure that forces firms to invest in innovation and to rapid dissemination and exchange of improved technologies throughout the economy. Besides competition there are five important factors that may explain the economic growth induced by capitalism:

1. *Oligopolistic competition:* A few big companies dominate a particular market. In this case innovation replaces price as the competitive weapon to ensure continued innovative activities and growth.
2. *Routinization:* Managing innovative activities and thereby minimizing uncertainty supports continuous innovation and sustainable growth embedded in day-to-day activities.
3. *Productive entrepreneurship:* In free enterprises, productive innovation is more incentivized than destructive occupations.
4. *The rule of law:* Contracts and property laws encourage innovation.

5. *Technology selling and trading:* Firms disseminate their knowledge profitably through licensing, even with their competitors.

Even though capitalism encourages entrepreneurs to innovate and profit from their innovations, corruption may harm and destroy innovations. This can be observed in many countries where people are not promoted based on merit, but based on attitude and knowing the right people.

Even China has come a long way from the days of Maoism with its technology-destroying Cultural Revolution to its present form of central government. China already accepted patent rights in 2001 due to its membership in the World Trade Organization (WTO) and introduced new property rights in 2007.

3.3 Technology and Growth in Modern Times

For more than 200 years every generation has had the feeling that it is more dynamic than its preceding generation. But from the nineteenth century onwards the speed of diffusion of technology has not changed: It takes about 10 years on average to reach 80 % market penetration, even in modern times. Do not forget that a today's market is global and not local compared to the beginning of the twentieth century.

Nevertheless, science is pushing towards new frontiers, engineers are improving existing technologies, introducing new ones, and society is more receptive to new developments than ever before. Innovation is self-reinforcing (Smith, 1776): Graduates from universities handing over the newest ideas from academia to industry conduct 99 % of technology transfer.

However, one has to keep in mind that "controlled progress would be no progress" (von Hayek, 2003). As Baron von Munchhausen said: "You cannot pull yourself out of a swamp by your own hair." Hayek was convinced that big innovations are not the result of a planned approach. Most generic new designs have their origin in "tinkering around" or just trial and error, like evolution (Simon, 1996). Scientists like to understand, engineers like to act. Sometimes they have to act without too much understanding.

In ancient times, technology was simply a tool to survive. In Galileo's times, around 1650, this changed, as the telescope, and more so the microscope, helped to improve our scientific understanding of the world. Technology was more and more seen as tool to improve the conditions of life. This culminated in the technocracy of the Ecole normale superieure in Paris: Auguste Comte (1798–1857) tried to apply natural sciences to social sciences; everything was within the grasp of the human brain. From the middle of the nineteenth century onwards, technology became more and more driven by science and today our whole society is very much driven by technology: Progress has become technical progress, a critical situation. Neil Postman, an American scholar of communication, calls it "Technopoly", the submission of all forms of cultural life to the sovereignty of technique and technology (Fig. 3.5).

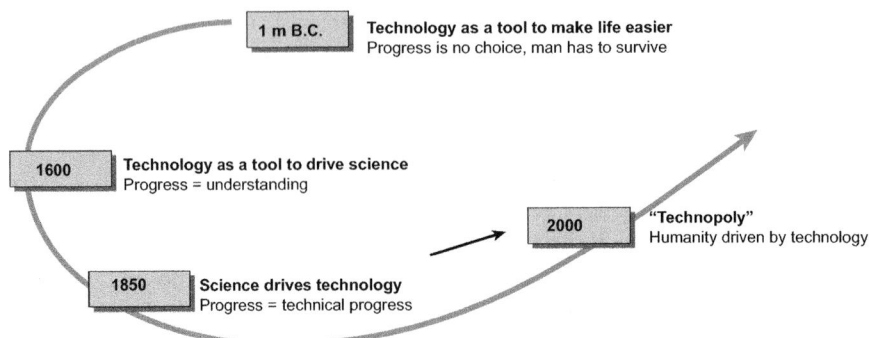

Fig. 3.5 Technology and the idea of progress go in parallel

Today we have an abundance of technical possibilities (Wilson, 1999). Energy may be produced with the help of rivers, windmills, coal-fired power stations, nuclear power, photovoltaics, deep thermal energy or wave energy. And in genetics we will be able to bypass natural selection not only in plants, but even human beings, for better or for worse.

More science means more responsibility: Technical awakening is a moral awakening as well. We have ever-higher aspirations and are able to trace minute effects and side-effects of every new and old technology. This leads the public asking for more remedies against long-term ill effects. Progress needs a new definition, modern technology makes too many things possible, and we are forced to make a non-technical choice. This selection may be different from region to region: "The most profound intellectual and ethical choice, humanity has ever faced." (Wilson, 1999). In rich countries we want to mitigate climate change whatever the cost may be. In developing countries, everybody wants to lead a decent life; adaptation is the preferred strategy to cope with climate change.

Whichever technologies will be used in the future is increasingly being decided by politics and not by the natural sciences. But many politicians do not understand the subtleties of technology and many engineers cannot follow political decisions. This is a dilemma we will have to learn to live with.

The Economist: The growth machine

Economists have thought too little about how innovation happens

© The Economist Newspaper Limited, London (May 16th 2002)

INNOVATION is an obsession of management gurus. Books such as "The Innovator's Dilemma" and "Innovate or Die" often top the business bestseller lists. That this is not an obsession shared by economists is remarkable. The dismal science certainly acknowledges the crucial role played by innovation in driving economic growth. But the fact that innovation happens is mostly taken as a given; the how and why of innovation remain largely ignored.

Some of the most up-to-date thinking on the subject remains that of Joseph Schumpeter in the first half of the 20th century. Indeed, the late great Austrian economist was lauded as a prophet of the "new economy" that briefly bloomed during the late 1990s, largely for his description of a process of "creative destruction" in which innovative new firms drive lumbering established companies from their dominant market positions. Never mind that much of Schumpeter's work actually celebrated the role of established firms in innovation.

To put right the previous oversight, William Baumol, a veteran economist at Princeton University, has written a splendid new book, "The Free-Market Innovation Machine"*. Building on the insights of Schumpeter and even of Marx (though this is a devoutly pro-capitalist tome), he argues that it is, above all, the ability to produce a continuous stream of successful innovations that makes capitalism the best economic system yet for generating growth. This is for two main reasons. Under capitalism, "innovative activity—which in other types of economy is fortuitous and optional—becomes mandatory, a life-and-death matter for the firm." Second, new technology is spread much faster because, under capitalism, it can pay for innovators to share their knowledge and "time is money".

If the capitalist system can be seen as a "machine whose primary product is economic growth", what are the machine's components? The rule of law, of course, especially the protection of property (including intellectual property) and the enforceability of contracts. This motivates innovators by ensuring that they can gain some reward for their efforts.

Some economists debate why countries such as America are more blessed with entrepreneurs than others (the former Soviet Union, say). Mr Baumol argues that there are entrepreneurs in every sort of system: what differs is whether or not they devote their energies to producing innovations that add to economic growth. And this depends on the incentives provided by the economic system. Non-capitalist systems often give entrepreneurs most of their rewards for innovations that do not contribute to growth, such as finding clever ways to win patronage from the state, creating monopolies, or crime. There is some of that in capitalist countries, but far more upside for growth-producing innovators.

From garage to lab

What of the lone entrepreneur building some world-changing invention in his garage? Such pioneers play an important role, particularly in devising radically new technologies. But a more important role in innovation is played by large established firms. They have changed innovation from an erratic "Eureka! I've found it" process into a more regular and predictable activity. Large firms with the resources to invest in "routinised" innovation and the incentive to do so are a unique feature of capitalist economies. Other economic systems have been innovative enough, but have failed to exploit their innovations—or have had big firms but no incentives to innovate.

The industrial structure that fosters productive innovation best is oligopoly, argues Mr Baumol. Oligopoly falls between monopoly—where one firm rules—and perfect competition, in which many firms compete and no single firm sets prices. In oligopoly, a few big firms compete with each other, but not primarily by trying to charge the lowest prices, which are thus usually higher than in a perfectly competitive market. Instead, oligopolists compete by making their products differ slightly from their rivals'. Innovation is a growing source of such product differentiation, says Mr Baumol.

It is ironic that competition through innovation by oligopolists may be the main driving force of growth and higher living standards, since oligopolies are often seen as a threat to the public interest and as a target for antitrust action. Mr Baumol urges antitrust authorities to judge innovative oligopolists less harshly.

Mr Baumol also tackles two other misleading beliefs. One is that innovators jealously guard their proprietary technologies through patents, lawsuits and secrecy, to maximise for as long as possible the above-normal profits they can earn from their innovation. In the real world, innovative firms are often remarkably quick to license new technology or to become members of technology-sharing consortia. Capitalist incentives explain why, says Mr Baumol. If other firms expect to be more efficient at exploiting your innovation, and so will pay more to use it than the innovative firm could make by keeping it to itself, it makes economic sense to license it to them.

Some economists argue that a shortfall in the legal protection granted to innovators means there is not enough innovation. Mr Baumol reckons that, on average, less than 20% of the total economic benefits of innovations go to those who invest directly or indirectly in making them happen. The rest of the benefit spills over to society at large. Arguably, stronger patents and other intellectual-property rights that made it easier for innovators to keep their proprietary technology to themselves or charge more for licensing it would increase total innovation.

But would that be progress? Mr Baumol doubts it. The rapid dissemination of innovation through the economy via spill-over effects has a hugely positive impact on economic growth, he says. Better protection for innovators might increase innovation but, by slowing its spread, reduce growth. For all the benefits that flow from innovation, it seems you can have too much of a good thing.

* "The Free-Market Innovation Machine. Analysing the Growth Miracle of Capitalism", by William Baumol. Princeton University Press

References

Baumol, W. J. (2002). *The free market innovation machine: Analyzing the growth miracle of capitalism*. Princeton, NJ: Princeton University Press.
Bils, M., & Klenow, P. J. (2000). Does schooling cause growth? *American Economic Review, 90*, 1160–1183.
Easterly, W., & Easterly, W. R. (2001). *The elusive quest for growth: Economists' adventures and misadventures in the tropics*. Boston, MA: MIT Press.
Samuelson, P., & Nordhaus, W. (2005). *Economics* (18th ed.). New York, NY: McGraw-Hill.
Schumpeter, J. (1942). *Sozialismus, Kapitalismus und Demokratie*. München, Bavaria.
Simon, H. A. (1996). *The sciences of the artificial*. Cambridge, MA: MIT Press.
Smith, A. (1776). *The wealth of nations* (A. Skinner Ed.). London, UK: Penguin.

References

The Economist. (2013, June 1). Poverty: Not always with us. *The Economist*, pp. 21–22.
von Hayek, F. A. (2003). *Recht, Gesetz und Freiheit*. Tübingen, Baden-Württemberg: Mohr Siebeck.
Wilson, E. (1999). *Consilience: The unity of knowledge*. New York, NY: Vintage Books.

Technology Speed and Substitution

> *We live in an age in which technological change is pulsating ever faster, causing waves that spread outward towards all industries.*
>
> Andy Groove, cofounder of Intel

Abstract

Most people think that the speed of technology change is accelerating. This is true for some specific technologies, but not for mature ones. Technology speed is a big challenge for all companies, especially when customers need changes fast and over-engineering looms large.

4.1 The Myth of the Increasing Technology Speed

Andrew Grove, one of the founders of INTEL, is convinced that his industry is moving ever faster (Grove, 1996). He is not the only one with this opinion. During the course of the last three centuries and possibly even in the centuries before, every generation believed that it was moving faster than the preceding one. At the beginning of the nineteenth century British textile machine producers could not sell their 3-year-old machines because they were already technologically obsolete. However, Grove is right: Some products in the electronics and software industry change very fast and special managerial attention is needed. Sometimes changes in regulation push markets in unexpected new directions: For many years the light bulb industry was only making minor improvements, but with the advent of LED technology together with CO_2 regulations it has regained its momentum (Bresnahan, 1997). Improvements in railway travel time, on the other hand, have a slow technology speed (Fig. 4.1).

Paul Krugman, an American economist and Nobel Prize Winner of 2008, is one of the few scientists who believe that an increasing technology speed is just a mirage: "We have no break-through innovations!" He compares today's kitchen with our kitchens of the 1950s: Not too much has changed. Most equipment we use today was already in place 50 years ago. One example is the microwave oven. The microwave is based on a patent from 1937 by Bell Telephone Laboratories and one from 1945 by Raytheon which describes the cooking process. The first microwave

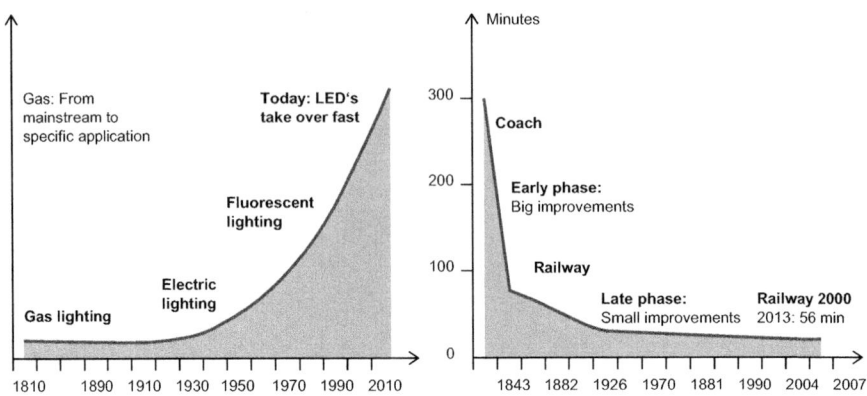

Fig. 4.1 On the *left* fast technology speed: better and better improvements in lighting since 1800. On the *right* slow technology speed: railway Zurich – Bern 1843–2013

was some 1.8 m tall, weighed 340 kg, cost about USD 5,000 and consumed 3 kW. Sales volumes of 40,000 units for the US in 1970 grew to one million by 1975. In 1985 roughly 25 % of US households owned a microwave oven. Krugman is right. But market penetration and prices have changed significantly since the 1950s.

Some examples from the sports industry seem to confirm Krugman's opinion. These examples show no improvement at the system level. For example, for the last 40 years, the winners of the ski race at the Lauberhorn in Wengen, Switzerland, have all completed the race in approximately the same time (Fig. 4.2). There have been no significant increases in speed.

However, the race itself has changed: there are more than 60 regulations in place nowadays. These regulations inspire technical innovation, but these innovations do not lead to a higher speed, to a higher overall performance of the system. On the contrary, the speed of the racers has been deliberately reduced. The slope has been changed. There are two reasons for these developments: First, being too fast can lead to dangerous and even fatal accidents, and second, technical innovation would give one racer too much of an advantage over all the other racers. This would lead to a monopoly: If the same athlete always won, the TV viewer would become bored, resulting in bad publicity for the races. The more regulations in place, the more random effects such as weather, wind, the momentary mood of the racers and the start number influence the ranking. And the more surprise there is, the more spectators are attracted.

Technology speed is a diffuse concept and therefore opinions diverge considerably. Once we choose a definition that allows measurements and quantitative data, we have to accept that technology speed is relative and that it depends on the goal we want to achieve (Rohner, 2010): For example, engineer M wants to design a platform for mobiles to integrate the most powerful camera modules available on the market. M is convinced that customers want to have as many pixels as possible

4.1 The Myth of the Increasing Technology Speed

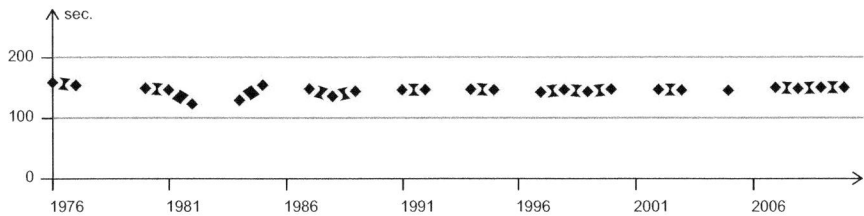

Fig. 4.2 Downhill race Lauberhorn: constant winner times for many years (http://www.lauberhorn.ch)

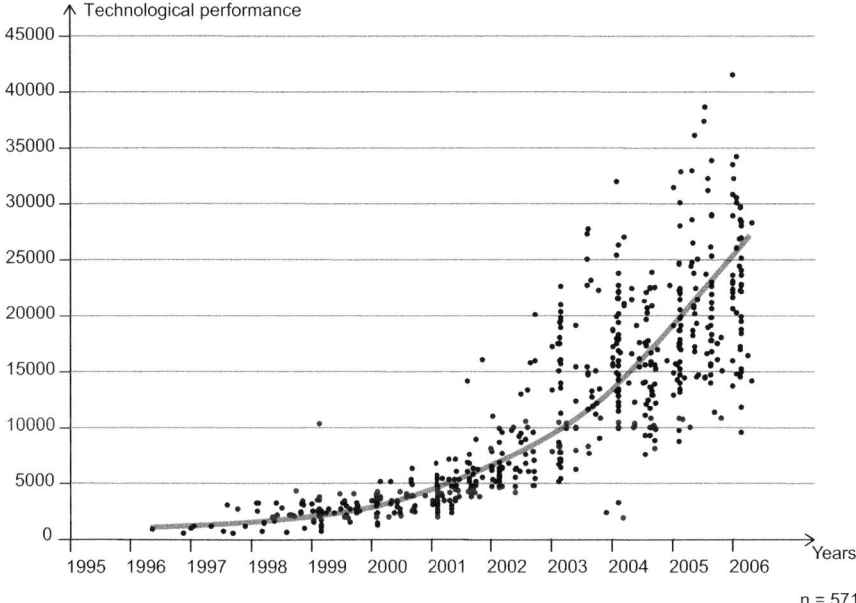

Fig. 4.3 Engineer M designs a platform to ease the integration of technological developments

in their camera. Thus he conducts a survey of all sensors available and discovers that technology speed is accelerating (Fig. 4.3): You get ever more pixels for your dollar. Thus a platform or modular design makes sense to allow for the sensor integration at a late stage in the product development process only. He will tell the logistics department not to stock any sensors; they will be obsolete in a short time. For him technology speed is defined as how fast the pixels per dollar ratio grows.

A second engineer W is brooding over the same problem. But he is convinced that 2 million pixels are good enough, and customers do not care about a few dollars. W carries out a market survey with the same data (!) as engineer M. She finds out that technology speed is decreasing. For her, it is important to get a fixed number of pixels per dollar (Fig. 4.4). This is her definition of technology speed.

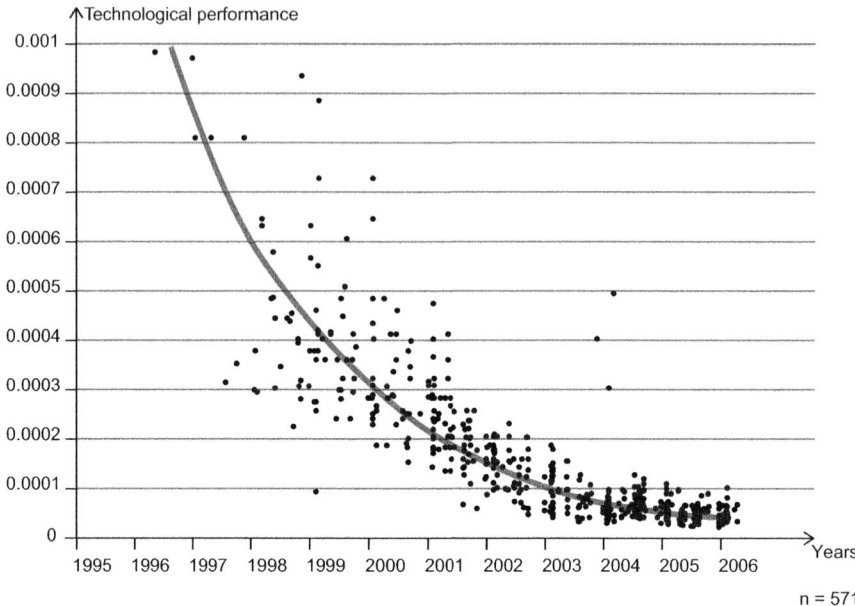

Fig. 4.4 Engineer W seeks other characteristics to differentiate her camera

She will tell the logistics department to stock sensors, because even second or third generation old sensors are good enough and costs will not change much.

We have to be very careful with technology speed. It tells us how to integrate modules into a product and it has far-reaching consequences for our supply chain. But it depends on what we want to do, on the goals we want to achieve for our customers. It is the speed we have in reducing the "distance" to this goal.

When a new technology enters the market, functionalities will remain constant in the first phase and the shape of the products remains unchanged as well. When the car replaced the horse, cars looked very much like an old coach. The coachman could keep his position; the shape of the coach remained constant, only an engine replaced the horse. This helped customers gain acceptance of the car. Old investments made on behalf of the coach were not made obsolete: The owner was not forced to buy a new garage. Car producers tried to make the change as smooth as possible for new owners. In the beginning new cars were not more than a gimmick for the rich: Expensive, unreliable and due to a lack of infrastructure only useful for short distances. The one big perceived advantage of cars around 1900 was the riddance of horse droppings, a big problem in big cities like New York. However, even some decades after the market introduction of cars, the horse was still seen as more reliable and more efficient. That can be seen by the number of horses killed during World War I: In 1917 Britain had over one million horses and mules in service. It lost close to 500,000 – one horse for every two men in the course of World War I.

Sports technology also shows that disruption and substitutions cannot be predicted: Nobody expected the material of the pole vault to change from wood to bamboo to metal and finally to fibreglass. Every substitution changed the nature of the game and made old technology and old styles of jumping obsolete. In contrast to downhill skiing, pole vault has only a few rules. It needs years of intensive training to master and has well-defined risks: The risk of falling some 6 m if you belong to the elite. The jump height of the winners increased from 3.6 to 6 m over a time span of 90 years. This is an improvement of 0.6 % per year, not an improvement as with CPUs, but much more than for the downhill racers at the Lauberhorn.

Pole vault shows that the timing of technology breakthroughs is difficult to predict: The change from bamboo to metal or from metal to fibreglass could be expected, but nobody was able to make a reasonable forecast as to when these substitutions would take place and nobody knows what the next big innovation will be.

4.2 The S-Curve

The popular model for technology substitutions is the S-curve: A technology starts off with small improvements, gains momentum, speeds up and towards the end of its life cycle it has diminishing returns, it needs more and more R&D for even minor improvements. New technology starts at a lower level than the existing one, but has a higher potential and therefore overtakes the old technology after a given time.

Currently car manufacturers spend approximately USD 10 billion per year on small improvements on the existing engine technology. Only USD 2–3 billion per year is invested in the new fuel cell technology, but this is improving several fold. Whether the fuel cell will overtake the existing diesel technology is thus only a matter of time, providing that the fuel cell and its predecessor, diesel technology, are measured with new criteria, e.g. increased environmental friendliness (Fig. 4.5).

S-curves are nice theoretical models, but useless for technology forecasts. However, we are deeply impressed by its simplicity and logic. A typical example is our recent discussion about peak oil. For years we have predicted that more and more countries are reaching their limits of oil production, such as the US. But US oil production is still on the rise. New technologies, such as hydraulic fracturing, often called fracking, become efficient once the oil price is high enough and big oil companies are convinced that the price will stay high for years to come. Fracking – injecting high-pressure fluid into rocks deep underground, inducing the release of fossil fuels – is an impressive technology that has been known for decades. But it is also a technology that imposes large costs on the public. We know that it leads to toxic waste-water that may, if not handled with care, contaminate drinking water. There is reason to suspect, despite industry denials, that some groundwater was contaminated at the beginning of its use. On the other hand, thanks to fracking the US has made the biggest contribution to lower CO_2 output worldwide over the last years by replacing coal with gas. Most probably we will not see radical changes with fracking as in the past when Edison had to battle against gas lamp

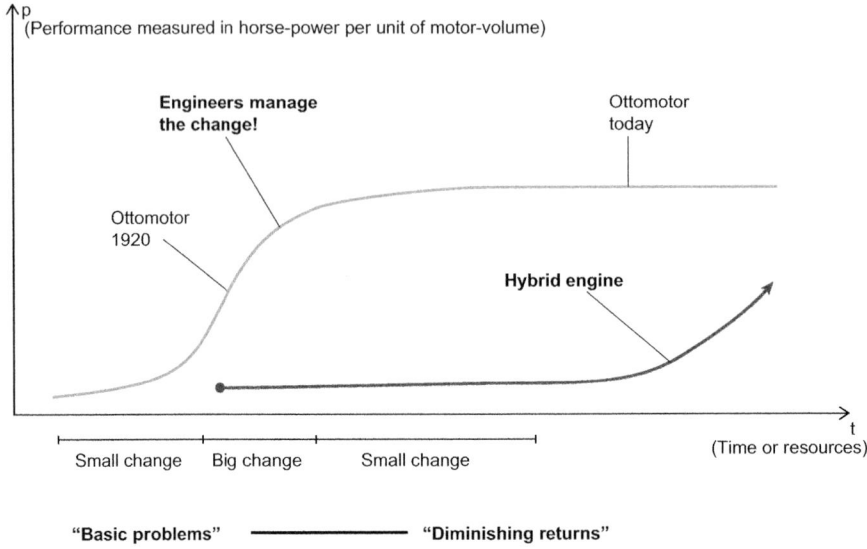

Fig. 4.5 S-curve: no problem for management, but too theoretical (Christensen, 1997)

improvements. But it may be that oil and gas will remain one of the most important energy suppliers for decades to come: Natural oil wells, when they are abandoned today, still have around 70 % of total oil potential left. Thus new technologies like fracking may have a considerable effect in future. Nevertheless, the one significant uncertainty about oil is not technology, but politics: Nobody knows how long customers will accept the CO_2 burden of oil. And neither does anyone know which S-curve will replace today's oil S-curve nor at what time this will happen. Regulations driven by social developments could change the acceptance of oil technology overnight.

4.3   Increasing Returns

The S-curve is too simple. Reality is quite different. We need additional information when we have to make decisions in a company about substituting technology. Why?

The first reason is that the growing number of end-users in modern technologies sometimes leads to "increasing returns" or better, increasing utility. For example, the 100th user of Lotus Notes gets a little bit more for his money than the 99th because he can get access to one more user. These increasing returns can often be seen in modern Information Technology (IT). The more users a specific software application has, the more R&D can be spent on its improvement, the more people using it for very special applications, the better the debugging works. Since software has low manufacturing and distribution costs, a steep volume effect occurs with an increasing number of users. Costs per user go down.

There is a second reason why the simple S-curve model does not represent reality. Whenever an old technology comes under pressure, it improves far more and much faster than expected. When Edison introduced his electric light bulb, the gas lamp improved its yield by a factor of 5 within a few years and lowered its costs by a factor of 3. The manufacturers of electric light bulbs had to revise their business plans and catch up with an improvement by a factor of 15. The same has happened several times with the fuel cell over the last few years. Old technology fights back; new technology fights a moving target. This is one explanation for the fact that most new technologies take much longer to get a decent market share than expected. Exceptions are technologies that provide a functionality never seen before. SMSs could not be installed efficiently on a telephone with a turning wheel. A second exception is sometimes due to new regulations: LEDs have to be bought even if they have much higher prices than the old light bulb!

Even the extended standard model, a more sophisticated model than the S-curve, has no predictive power. Its components innovation, communication channels, communication over time and the social system can be used to make descriptions at most. The model shows which parameters could be decisive. Kahneman (2011) is even convinced that sophisticated models are worse in complex environments, since they only describe correlations. He proposes to use simple models. Once a model contains many variables it can easily be used to map history. But then we run the risk of seeing the future as an extrapolation of the past, which is for technologies a rather strange assumption!

4.4 Acceptance of New Technologies

Finally, new technologies have to be accepted: Customers, end-users, manufacturers and retailers all have vested interests. They are usually not risk-averse, but want to protect their investments.

What then makes people accept new technologies? One decisive factor is the impact on end-users and employees in the value chain. If the new technology is competence-enhancing for employees, they will accept it easily. For example, going from milling to high-speed milling is not difficult for the shop-floor employees. The change from metal bashing to plastic injection moulding destroys the competencies of old employees and will be resisted.

Nobody likes to have invested in vain, be it lost investments in training or in machines. And most people do not like to accept that younger employees are better prepared for new technologies despite having spent much less time for building up competencies.

4.5 Disruptive Innovations

The biggest problems arise with disruptive innovations, innovations that may serve "lower markets" or less demanding customers at the beginning but have the potential to become "good enough", as good as the old applications and be able to fulfil the needs the end user has. In the beginning, disruptive innovations do not represent competition for traditional technologies, but they
- are more expensive
- emerge in small, insignificant markets (niches)
- have lower gross margins
- take away least profitable markets
- have lower technical performance (measured with the metrics of the old needs)
- are perceived to be better in a niche with new metrics, new needs.

If a new technology emerges, established firms and new entrants use different strategies: While new entrants push into new markets where the old technology is either an overkill or not applicable, established firms experience a market pull from their existing markets. As soon as new firms gain more experience and improve their efficiency, they can gain access to big established markets and threaten existing and successful firms (Christensen, 1997). Even though Peter F. Drucker said to "watch the non-customers", successful companies face the dilemma of looking for profit in existing mature big markets and neglecting new markets or going for new markets and losing profit. Listed companies sometimes have the problem that entering new markets with lower profits will make them less attractive for financial investors.

One example is the development of the hybrid car: Volkswagen concentrated on offering the same Golf model to its customers, as it found that existing customers just required some more speed for less money with a little more performance. In contrast, Toyota introduced the first Prius in 1997 in Japan. It was the first mass-produced hybrid vehicle. As a newcomer in hybrid technology, it had hardly any competition at the beginning and incumbents did not defend the new market. It was too small and profits too low. Toyota became more and more experienced and in 2001 took over new markets by introducing the Prius worldwide. In 2004, Lexus, the luxury division of Toyota, introduced the first Lexus hybrid with higher margins and even less competition.

Hard disks also provide a good example: For years, a 14 in. diameter was the standard. Engineers improved the disks from year to year. When the first 8 in. drives reached the market, "14 inch engineers" understood the new technology very well although it did not make much sense to them. In their view, big hard disks were much more useful and had much more potential – seemingly forever – than the small disk technology. Eight-inch technology would nullify a lot of the investments the company made in their manufacturing sites and in their customers' applications. Most customers were asking for even more memory space, a need that only could be filled with 14-in. technology. Thus the market leaders pushed the 14 in. and the new 8-in. technology had no chance in the main market, only in niches where size was a decisive factor, performance was low and margins were small. But the

4.5 Disruptive Innovations

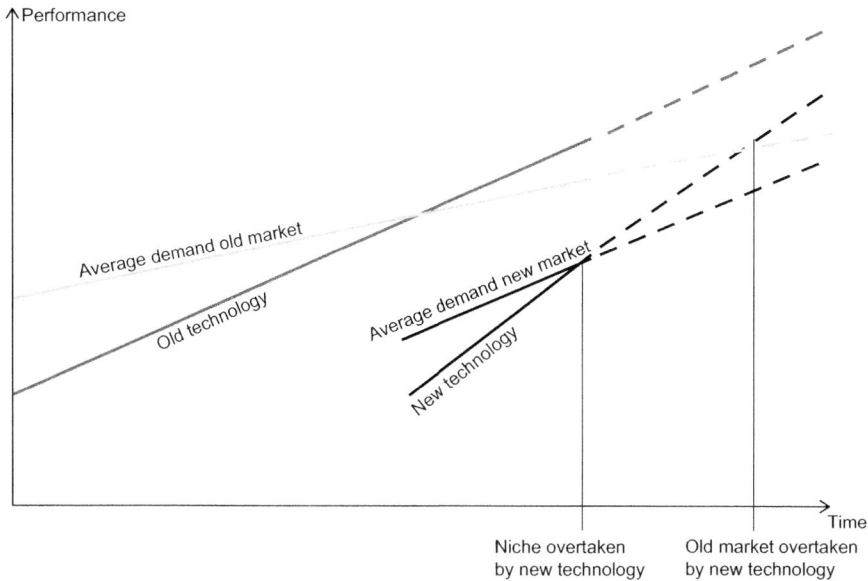

Fig. 4.6 New technology starts in niches, takes over larger markets (Adapted from Christensen, 1997)

average end-user was more and more pleased with the improved storage capacity of the 8-in. technology. Fourteen inches became an increasingly over-engineered technology. It lost market share, was pushed into a niche and the smaller disk overtook the big disk within a few years (Christensen, 1997). The smaller disk was not only smaller, but also cheaper. The old leading companies had no chance, were not able to pick up and took a back seat. The old winners lost because they had been too successful and too customer-minded. This is the interpretation of MIT scholar Christensen (1997).

The dilemma of successful leaders – pushing their successful technology and losing in the end to a less powerful technology – has been observed in many markets. It seems to be extremely difficult to keep a successful technology long enough and to change the technology at the right moment. New technology starts in niches and can take over larger markets (Fig. 4.6).

An impressive example is Kodak, best known for its wide range of photographic film products. During most of the twentieth century Kodak held a dominant position in photographic film. The company's dominance was such that its tagline "Kodak moment" entered common dictionaries. But in the late 1990s Kodak struggled financially as a result of the decline in sales of photographic film. As part of its turnaround strategy in 2007, Kodak focused on digital photography and digital printing and attempted to generate revenues through aggressive patent litigation. On January 19, 2012, Kodak filed for bankruptcy protection and obtained a USD 950 million, 18 month credit from Citigroup to enable it to continue trading. On February 9, 2012 Kodak announced that it would stop making digital cameras, pocket video cameras and digital picture frames in the first half of 2012. Kodak intends to focus on photo printing, through retail and online services as well as desktop inkjet devices and will license its brand name to other camera manufacturers.

In mature companies there is always a better project with the existing proven technology than with a new emerging technology. Mature markets are big, much bigger than emerging markets! New technologies start small and at the beginning of their life cycle cannot solve the shortcomings of the old technologies. And since they serve small markets only, they cannot solve their companies' big problems.

Initial potential protection against such dilemmas is investment in process technology and process innovations. Both lower costs and make it difficult for a competitor to start a price war. Process innovations in lighting, which started after the biggest product innovations in 1885, made light bulbs affordable for the mass market.

On the other hand, big investments in processes for the old technology make it even more difficult to change at the right moment. Therefore technologies under pressure try to open new markets or niches, as in the case of gas technology. From 1800 to 1880 gas was distilled from wood and coal, and most gas companies held a monopoly position, as did the first gas company in the world in 1812, the Gas Light and Coke Company in London. Problems with escaping gas and explosions led to the rapid adoption of electric light and gas illumination came under pressure. With big improvements, gas lighting could defend its competitive position for some years and has not been totally replaced today. It has been pushed into niches like camping or houses far away from electrical grids, such as lodges during the winter in the north of Norway.

However, top management does not easily accept writing off sunk costs. Companies like Gillette use a secondary line of protection against the dilemma of disruptive innovations: They practise cannibalism and outpacing: They replace their successful products when they are still thriving in order to ramp up prices for their new products and move faster than their competitors. And they are making good profits. To innovate in such an aggressive way, they spend more than 2.2 % of their sales in R&D. Much of its R&D goes into manufacturing processes: They invest heavily in dedicated machinery.

The Economist: Frugal innovation: Green ink

© The Economist Newspaper Limited, London (Mar 1st 2012)

BY 2017 printing presses around the world will lap up 3.7 m tonnes of ink, worth some $18 billion. Most of it will contain hydrocarbon-based solvents resulting in emission of volatile organic compounds (VOCs), an undesirable by-product of the manufacturing process. But not all. EnNatura, a company spun out of the Indian Institute of Technology in Delhi has created a formula for making ink that is environmentally friendly.

Printing ink is made using four components: resins, solvents, pigments and additives. Resin binds the ink together into a film that sticks on the printed surface. Solvents ensure consistent ink flow. Pigments give ink its colour and additives control its thickness. Raw resin, however, is no good. To make it suitable for printing, it is heated in a closed kettle along with a variety of ingredients at high temperatures. This mixture is spiked with petroleum distillates to tweak the resin's molecular structure such that its bonds are easily breakable. EnNatura's proprietary resin chemistry does the same thing using castor oil, a natural purgative.

Resin is then mixed with a vegetable-oil-based solvent to make a thick paste called varnish. The solvent ensures that the ink does not dry too soon (while still on the printing plate) or too late (once applied on the paper). In conventional ink manufacturing, hydrocarbons like ethyl acetate and isopropanol (rubbing alcohol) are added for better solubility and to hasten the drying process. As EnNatura's resin sports a loose framework of molecules, it mixes readily with linseed oil, a natural solvent and an effective drying oil, explains Krishna Gopal Singh, the company's co-founder.

Other companies, especially in America, make biodegradable ink. But most use petrochemicals to clean the resin from printing plates once the printing job is complete, which defeats the purpose, according to Mr Singh. EnNatura, by contrast, employs a liquid concentrate made from a surfactant, a substance which, when mixed with water, eats into the resin and scrapes it off the printer.

A lithographic printer typically uses half a litre of liquid in the washing process. An eco-friendly wash in America or Europe costs around $10 a litre whereas a hydrocarbon substitute goes for about $3. EnNatura offers its non-polluting solution for just $1 per litre. (However, its ink sells at about 7-10% more than its conventional counterpart. The difference is due to the higher cost of vegetable oil as compared to a petroleum-based one.)

Devoid of any hydrocarbon solvents, EnNatura's ink, called ClimaPrint, also poses no health hazards to shop-floor workers. This, however, does not seem to be a strong selling point. Most printers respond that they have yet to suffer from any adverse health effect, Mr Singh sighs. Such apathy can be attributed to lax enforcement of regulations against pollution in general. These can be bypassed by greasing the right palm. As a result, printing presses get away with dumping the after-wash waste in the nearest gutter. After much cajoling, though, a few local print-shop workers in South Delhi have now started to print brochures, leaflets, posters and limited-circulation magazines using ClimaPrint.

This small success has been a long time coming. EnNatura took five years to launch its first product, partly due to lack of experience in the ink-making business. Inadequate funds also meant that the laboratory equipment was not always up to snuff. They got around initial hitches by resorting to what Mr Singh likes to call "jugaad", or solving problems using only those resources, which are at hand. Thus a dough blender helped grind pigment into the varnish. Instead of using a viscometer, an instrument which measures viscosity of liquids and costs a hefty $20,000, Mr Singh turned to local mechanics who built a replica at less than one-hundredth that. Although readings taken from this clunky tool were not terribly accurate, it proved a great help nonetheless. "In the early days, our ink was way off when compared with conventional ink, so a few percent here and there in the reading didn't matter," he recounts.

The 15-member team (excluding the two founders) is made up of graduates with non-engineering backgrounds, mainly because engineers demand far higher salaries. Despite striving to cut costs, Mr Singh jokingly feigns indignation at having to spend 250,000 rupees ($5000) for a lawyer to write the patent document to shield a young start-up from bigger companies.

One such is Japan's DIC Global, the world's leading ink manufacturer with a very different approach to tackling the VOC-emissions problem. It harnesses waterless lithography, which, unlike offset printing, uses a layer of silicon on the printing plate instead of water to keep the non-image areas free of ink. Thus water, which is used in huge quantities in offset printing, is spared. It also relies on an imaging technology called computer-to-plate in which image is transferred onto to the printing plate digitally. This has replaced the older computer-to-film procedure where images are first transferred onto a photographic film, which entails the use of chemicals for photo processing. In addition, the ink can be easily washed off using water rather than petroleum based solutions.

This high-tech solution is gradually gaining ground, but it requires a switch to advanced waterless presses. EnNatura, which hopes to license its ClimaPrint formula to ink-manufacturers outside India, improvises with the existing infrastructure. For now, then, its model has promise printed all over it.

References

Bresnahan, T. F. (1997). *The economics of new goods*. Chicago, CA: University of Chicago Press.
Christensen, C. (1997). *The innovator's dilemma: When new technologies cause great firms to fail*. Boston, MA: Harvard Business Press.
Grove, A. S. (1996). *Only the paranoid survive: How to exploit the crisis points that challenge every company*. New York, NY: Broadway Business.
Kahneman, D. (2011). *Thinking fast and slow*. New York, NY: Farrar Strauss & Giroux.
Rohner, N. (2010). *Technologiegeschwindigkeit*. Dissertation, ETH Zürich.

Diffusion of New Technologies

> *Diffusion is the process by which an innovation is communicated through certain channels over time and among the members of a social system.*
>
> Rogers, 2003

Abstract

Diffusion of new technology usually takes 10–40 years to achieve the big majority in a market. The first example studied, hybrid corn in Iowa, was S shaped. Since then all schoolbooks examples are S shaped. But reality is different. We often have a dip in sales after a first take-off that kills many small companies.

New technologies are accepted and spread because they solve a problem; they give its owner a perceived utility. This utility can be pushed by law, fashion, economies or justified by pure curiosity. Once a new technology is on the market, it is optimized and an average improvement rate of 3–4 % seems sustainable over longer time periods; once mature, improvements will go down to some 2 % per year. But there are exceptions! For a long time battery technology was more or less constant. Since we have come to need more and more power in mobile phones, capacity per 100 g has increased considerably.

Growth may follow many patterns, but the standard model is still the S-curve and goes back to the adoption of hybrid seed corn in Greene County, Iowa between 1930 and 1940.

Hybrid seed corn is no longer a product studied in management courses. In today's products we have more and more software. Additional functionality runs parallel with the growth of sales. Sales in new markets generally follow the hockey stick pattern: After the market launch, sales are very low for some time before they take off at the tipping point (Fig. 5.1).

When exactly sales take off depends on so many parameters that it cannot be predicted. The time to sales take-off depends mostly on industry and product type and finally on the strength of the company behind the new product (Fig. 5.2). At British Petroleum, central planning and corporate strategy used a simple rule of thumb: You need 10 years to get a market share of 1 billion dollars.

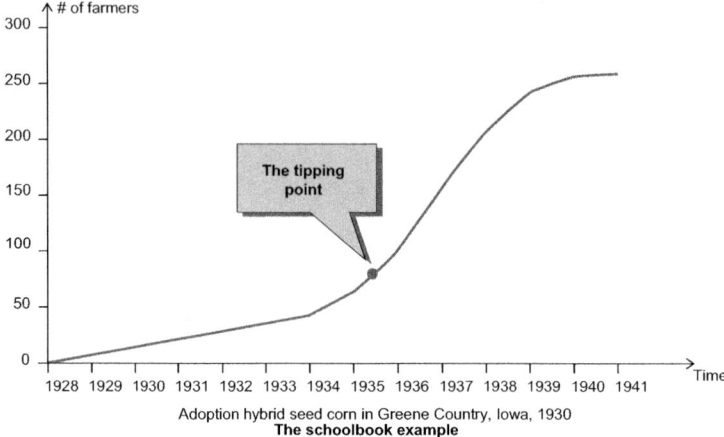

Fig. 5.1 New technologies spread in an S-curve, sometimes (Ryan & Gross, 1943)

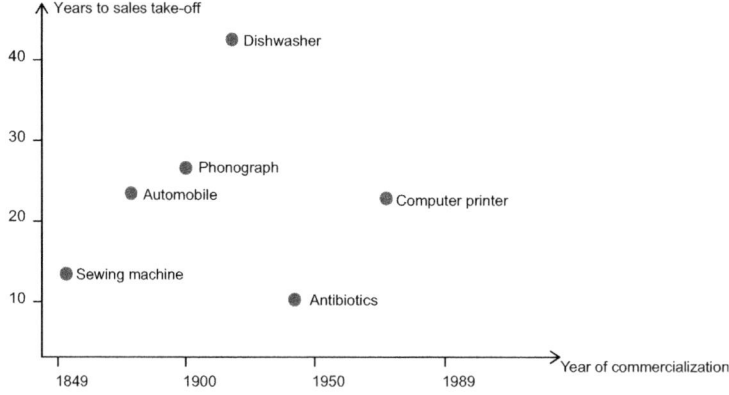

Fig. 5.2 Time to sales take-off depends mostly on industry and product type, and cannot be predicted

For some product innovations such as compact disc players, cellular telephones, radios, and jet engines, a large fraction of the competitors during the early years were new entrants. Stiff competition was the result. For other products such as dishwashers, electric blankets and vacuum cleaners, relatively few firms entered during the initial years of commercialization. Similarly, the rate of price declines varies significantly across markets, with some innovations actually achieving an increase in prices. For example, given the dramatic increases in the quality of turbo-jet engines, prices actually increased during the early years after market launch.

5.1 Diffusion of Innovation

According to Rogers' diffusion of innovation theory (Rogers, 1962), diffusion is a process by which *'an innovation is communicated through certain channels over time among the members of a social system'*. These four main elements are crucial:

1. To develop an innovation, Roger identifies six main steps: (1) recognizing a problem or need, (2) basic and applied research, (3) development, (4) commercialization, (5) diffusion and adoption, (6) consequences. Innovations that are perceived as having greater relative advantage, compatibility with customer needs, ease of trials by customers and observability of results and less complexity will be adopted more rapidly than others.
2. Given that an innovation has been implemented, communication must take place in order to share knowledge and information, which leads to diffusion.
3. The time in the diffusion process can be characterized by (1) the innovation-decision process, (2) the innovativeness of an individual component compared to the system, and (3) the innovation's rate of adoption in the system.
4. The social system – individuals, informal groups, corporations and NGOs – has a big influence especially on disputed technologies such as gene-tech, biotech, and nuclear power.

This model has been modified by several authors who argued that Rogers' theory cannot satisfactorily describe the diffusion of complex and networked technologies. Therefore several other studies have extended his model. But even those more sophisticated models have no predictive power and can only be used to make descriptions at most.

The study of many new products and technologies in the US shows that five parameters can explain many diffusion patterns in hindsight. Two parameters affect diffusion speed directly: Price and the number of firms entering the field. These two parameters depend on two more parameters each: The required R&D resources push up prices and reduce the number of entering firms. The entry of large companies reduces prices and finally the general level of innovative activity increases the number of new entrants (Bayus, Kang, & Agarwal, 2007) (Fig. 5.3).

The most important parameter is the innovative activity in the industry. It can be measured through the number of patents issued. The more patents we have, the more new firms enter the field. But it is easier to count patents than to count companies, and patents appear before new firms enter the field. As a rule we can say that the time to sales take-off is indirectly proportional to the number of patents. Rapid sales take-offs in new markets come mostly from innovative activity and the entry of large firms.

Required R&D resources seem to be a huge barrier for the entry of new firms and they push up prices on the market. A typical example is large pharmaceuticals with their large R&D costs.

In contrast, the entry of large firms decreases prices and also has a second effect: As soon as big companies enter the field, customers become more confident and buy the product. PCs became popular only after big IBM entered the field: A typical

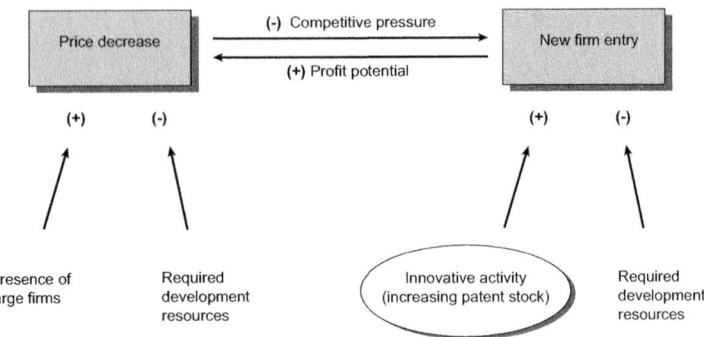

Fig. 5.3 A theoretical framework of diffusion: innovative activity is the key driver

example is the personal computer industry. Even though PCs were conceptualized in the 1950s, they only became technologically possible in the 1960s. The first commercial product, the Microcomputer, became available in 1974; the Altair was introduced in 1975. In 1974 there were only few companies selling a few thousand units, prices were high and sales dragged on for more than 5 years.

When the number of companies selling PCs grew slowly, prices plummeted by a factor of four, but sales remained flat for another 3 years. Then the tipping point came about with the arrival of the IBM PC in 1982: Sales rocketed, over 3 million units were sold – an increase of 291 % over the previous years. More companies entered the field and prices remained flat despite the huge improvements in memory power and calculation speed (Fig. 5.4).

The real breakthrough came when people recognized the flexibility of the PC: Prices were down and attractive software applications were on the market. Standards evolved through increasing returns. Today the PC seems to be over its zenith, as sales decreased for the first time in 2012. Price pressure from tablets is increasing.

When a product meets dormant customer expectations, growth reacts in direct proportion to offers. This was the case in the US cellular telecom market. This market created thousands of jobs with very short pay back periods from the beginning. This explains why companies were prepared to pay huge sums for the rights to set up a network in a new region.

The US was certainly the pioneer of mobile phones. The first license was given to AT&T in 1945. In 1956 a first prototype was installed in Saint Louis, but the first real system came into use in Chicago. It was an analogue system, able to cope with a few thousand users only and it was immediately sold out. The rest of the country was distributed through the Federal Communications Commission (FCC) to many different providers. US anti-trust authorities wanted to have at least two providers competing in each region. This prevented standardization until 1994 when through mergers the first true standard evolved and allowed an economy of scale similar to that in Europe. Competition was, in this case, not beneficial for the end-user, as everybody could confirm who travelled in the US in those times.

5.2 Standards

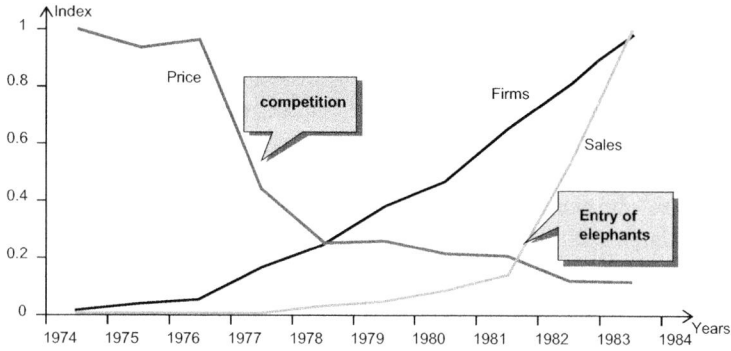

Fig. 5.4 PC Industry: driven by technical progress and trust (adapted to Bayus et al. 2007)

5.2 Standards

Even though Sweden launched the first mobile phone system in 1955 in Europe, the real start was in 1969 when engineers from the Nordic countries met and set up the first international standard, the Nordic Mobile Telephony (NMT). It was the first fully automatic cellular phone system and highly successful. In 1981 Sweden already had more than 20,000 users. In 1987, 5 years after the launch of NMT, 2 % of the population were subscribers. NMT was an analogue system, suited to the transmission of voice, but hardly anything else. NMT specifications were free and open, allowing many companies to produce NMT hardware and push the prices down. The success of NMT had a huge impact on Nokia and Ericsson. Initial NMT phones were designed to be mounted in the trunk of a car, with a keyboard and display unit at the driver's seat. "Portable" versions existed: One could definitely move them, but they were bulky, and battery life was a grave concern. Going digital created the opportunity to provide new services. More importantly, it created the chance to establish a European standard for prototypes in 1982 and in 1987 for a Global System for Mobile Communication (GSM), pushed by the European Union. GSM became an international standard within a few years. After the launch in 1991, Europe was covered and by 1996, 103 countries had adopted the standard. Europe had overtaken the US.

The mobile phone industry is a typical example of how standards evolve in open products, in products that involve many different parties, require heavy investment and infrastructure used by all participants. It is either a market leader taking advantage of network effects or a political will that triggers the standardization process. Mobile phones needed some 20 years to become standardized. Containerization shows as well that it may take 20–30 years to reach a global standard.

5.3 Industry Structure

The mobile phone industry is highly concentrated. The Top five mobile phone manufacturers in 2010 (Samsung, Nokia, LG Electronics, Zhong Xing Telecommunication Equipment Company Limited (ZTE), Apple) control more than 70 % of the world market. Sony Ericsson has been ousted from the Top 5 List for the first time since 2004. Instead, Apple has leapt ahead with the help of the iPhone 4. And in 2013 Nokia sold its hand-held device business to Microsoft. The market will remain heavily contested. It is a huge market with a big growth potential and technology in mobile phone components is moving fast.

In these highly competitive fields a further push can be expected from the convergence of markets and networks. Communication, consumer electronics and computers converge on the market side and different network standards will congregate on the technical side. This will change the industry structure as well.

In the past each provider of mobile phones had to integrate many different technologies and master mass production, integration, even the production of base stations. Only large integrated companies could manage this variety. However today, standards, miniaturization and the spread of know-how have led to a modularization of the total system. Consequently, the unbundling of the value chain will lead to a formation of specialists for modules and original design manufacturers (ODMs). The competition of vertically integrated companies will move to a competition of value chains and a horizontal competition of specialists where everybody is looking for economies of scale. This development is very similar to that of the PC business.

With technology moving from a bottleneck to a commodity, ODMs try everything to get new means of differentiation, which leads to new business models. As early as 2004 in Korea, users did not pay the use of network time, but the use of application time. It remains to be seen which companies are able to cope with these powerful changes.

5.4 The Saddle

When a new technology comes to the market, we have a pattern of generic phases which is followed in most cases (Fig. 5.5):
- Innovators: They are convinced that their technology is the best on earth.
- Early adopters: People who like playing around with new technology or have an urgent need. They have a vision similar to the innovators that is compatible with the new technology.
- Early majority: They wait and watch whether the new technology is successful.
- Late majority: They just follow the others. They usually optimize and cherry-pick.
- The laggards: At the very end you are pushed into the new technology because no other solutions are available anymore.

5.4 The Saddle

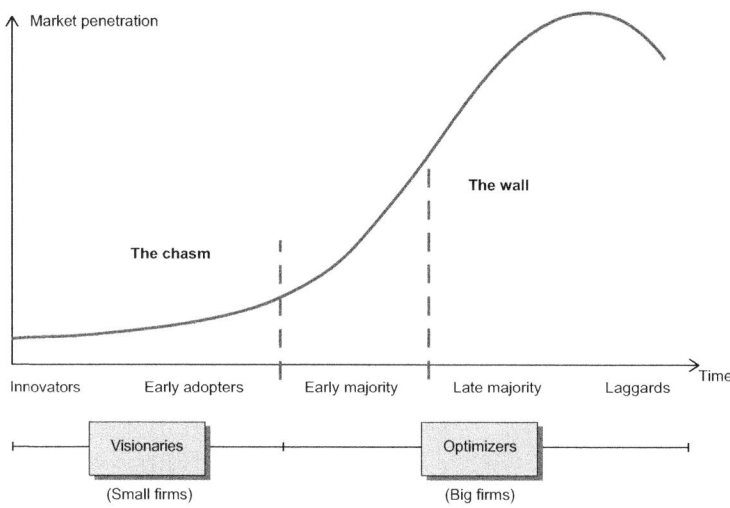

Fig. 5.5 Between early adoption and early majority: the chasm

Innovators and early adopters are in many cases small and medium enterprises (SMEs). In SMEs decisions are taken by the owner who in the end bears all the risks and does not need many employees to calculate whether a new technology has a sufficient payback period or not. The owners do not have to justify themselves in front of financial analysts either. Whenever they see an opportunity to solve an urgent problem, they take the chance.

Sometimes we have a dip in sales between the early market adopters and the early majority: The chasm. It takes time to convince bigger companies, and the markets for early small customers may already be in decline. During this phase we see a great difference between the intent to buy and actual buying. There was a dip of 30 % in the PC industry which lasted for 7 years. For mobile phones and VCRs the dip lasted 3 years and sales dropped by more than 30 %. This dip is responsible for many problems in start-up companies and is one of the reasons why venture capitalists sell their stakes as soon as the market takes off. The occurrence of chasms is neither a market failure nor the result of big companies being risk-averse. It is simply the overlay of the early market and the late market. If both are of similar size but have big time delays, the chasm opens up right after the culmination of the early market (Fig. 5.6).

According to the model of adopter categorization by Rogers (1962), innovators and early adopters are the visionaries, whereas early and late majorities and laggards are known as the optimizers. Most firms that start innovating or adopting early products are small, looking for opportunities and solutions to simple problems. The early or late majority, or even the laggards, are mostly big firms that need predictability and want stable products with a decent service.

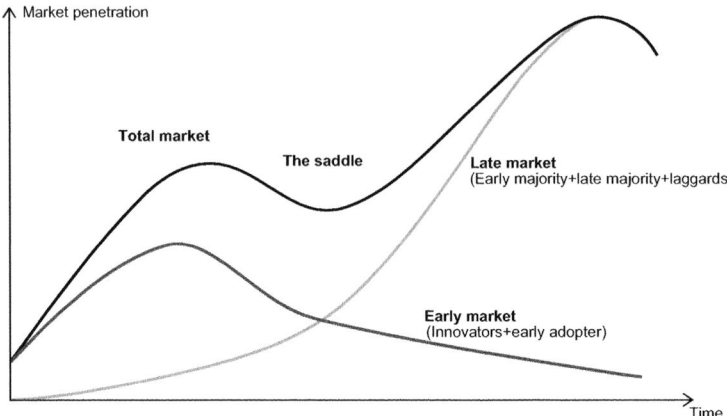

Fig. 5.6 The late market, such as early and late majority and laggards, profit from a recovery after the saddle

5.5 Technology Spreads Like Rumours

A Chinese teacher was travelling through Maine on vacation in the summer of 1945, shortly before Japan's surrender to the Allies at the end of World War II. The teacher was carrying a guidebook which said that a splendid view of the surrounding countryside could be seen from a local hilltop. He stopped in a small town to ask directions. From that innocent request a rumour quickly spread: A Japanese spy had gone up the hill to take pictures of the region (Gladwell, 2013).

Rumours are like news of new technologies (Allport, 1947): There is a fact that leads to a story, some details are left out, and then the story is tweaked to make sense in an abstract way. The story needs to be assimilated to make sense in the current environment and so the message sticks. Rumours spread with three specifics:
1. Little causes have big effects. There is just a small difference between stability and extreme growth.
2. Connectors, people who know everybody, have a big influence on public opinion. Venture capitalists usually ask well-known peers for their opinion. And "technology gurus" are able to influence the early adopters, less the early majority.
3. The message has to stick. It has to fit the environment in an abstract general way yet it has to be rather specific regarding the situation of the individual addressed.

The sheer complexity of markets, potential customers and the technology itself needs a simple message to convince the public: The value proposition has to be simple! Engineers do not like to pursue markets with non-technical messages. And most of them have learnt the hard way that details are important.

Mass-communication is different: simplicity, simplicity and then repeat and repeat and...!

The Economist: Fuel economy

Difference engine: Going along for the ride
© The Economist Newspaper Limited, London (Feb 3rd 2012)
IMAGINE you are stopped in the street by a clipboard-toting pollster, who asks whether health insurance should automatically cover all necessary procedures and medication, with no restrictions or co-payments? Nine out of ten people (if not all) would instantly answer yes. But had the respondents been warned beforehand that they would have to stump up an extra $10,000 for such coverage, the answer could easily have been a resounding no.

Sad to report, this kind of selective polling—in which only the benefits are mentioned so as achieve a desired result—is cropping up increasingly in the support of various agendas. Substitute motor manufacturers for health insurers and fuel efficiency for medical treatment in the thought experiment above, and you have the kind of "grassroots justification" that the Environmental Protection Agency (EPA) and the National Highway Traffic Safety Administration (NHTSA) are using to push through a doubling in the Corporate Average Fuel Economy (CAFE) figure motor manufacturers must achieve or face stiff penalties. The new proposal requires the fleet-average for new vehicles sold in America to rise incrementally from today's 27mpg (8.7 litres/100km) to 54.5 miles per gallon by 2025.

The most effective endorsement for the new CAFE proposal comes from a national survey carried out last October by *Consumer Reports* (a highly respected non-profit organisation based in New York) which claimed that 93% of the 1,008 people interviewed by telephone wanted higher fuel economy and would pay extra for it. The survey results were released the day before the government announced its new CAFE proposal. Ever since, the poll has been considered hard evidence that the new CAFE proposal is widely supported by the American people.

Like all polls, when people are asked personal questions, they do not necessarily tell the whole truth. Indeed, most respondents provide the kind of answers they consider socially acceptable. In some cases (how often do you brush your teeth?) the response is inflated; in others (how much do you drink?) it is deflated. It is a well-known phenomenon, called "social desirability response bias", which has been studied by sociologists and market researchers for decades and requires questionnaires to be structured in highly specific ways to minimise inbuilt biases.

"So, when consumers say fuel economy is extremely important, they are not lying," says Jeremy Anwyl, vice-chairman of Edmunds.com, the most popular source of independent car-buying advice used by American motorists. "The survey has just not allowed (or required) them to point out that other attributes are even more important."

Deciding what new car to purchase is a complicated undertaking, requiring all sorts of rational and emotional compromises. There are emotional factors to take into account, such as brand image, luxury, sportiness and appearance. Then there are physical features like comfort, number of seats, cargo space, body type and towing capacity. For most people, price and fuel economy represent costs.

In testifying before Congress last year, Mr Anwyl explained that "consumers are happy to pay less, or save fuel, but *not* if it means giving up features they deem important." A market simulation model that Edmunds has developed over the years indicates that fuel efficiency accounts on average for about 6% of the reason why consumers purchase a particular vehicle. Fuel efficiency only becomes important when petrol prices suddenly start climbing, but then declines in significance when pump prices stabilise or start to fall.

Here, then, is the conundrum: if the vast majority of motorists in America claim to want higher fuel economy and are prepared to pay for it, why are they not doing so? There is no shortage of cars already on the market capable of 40mpg or more. The Toyota Prius gets 50mpg on the EPA's combined cycle, while electric vehicles like the Chevrolet Volt and the Nissan Leaf deliver the equivalent of 60mpg and 99mpg respectively.

Yet few people are buying them. Despite large tax incentives for buyers, Volt and Leaf sales have fallen well short of even their manufacturers' modest expectations. Altogether, hybrid and electric vehicles account for only 2% of America's new car market. The government may mandate manufacturers to produce fuel-efficient vehicles, but it cannot mandate motorists to buy them.

The problem is their higher purchase price. Take the Toyota Camry, which is available in directly comparable hybrid and conventional forms. The hybrid version costs typically $3,400 (or 15%) more than the standard car, but achieves 13mpg (or 46%) better fuel economy. Even so, there have been few takers for the petrol sipper. Last year, Toyota sold over 313,000 ordinary versions of the Camry compared with fewer than 15,000 hybrid versions, notes Mr Anwyl.

The official line is that owners recover the premium on a hybrid or electric vehicle through lower running costs—with payback times of anything from four to seven years, depending on fuel prices. Unfortunately, American motorists tend to trade their new vehicles in after 3–4 years, and so do not collect anything like the full payback. Besides, the Edmunds model of consumer-buying habits (with 18m visitors a month, the website collects more data on such transactions than even the manufacturers) shows that consumers demand a payback period of 12 months or less.

What, then, to make of the new CAFE proposal? Achieving a fleet-average of 54.5mpg is doable, say motor manufacturers, but it will not be cheap. The NHTSA puts the cost of the extra technology (eight-speed transmissions, direct-fuel injection, turbo-charging, hybrid drives, larger batteries, electric power steering, low-rolling-resistance tyres, better aerodynamics, heat-reflecting paint and solar panels) needed to achieve the target at a modest $2,000 per vehicle, with a lifetime saving on fuel costs of $6,600 (provided fuel prices do not decline).

At the other extreme, the National Automobile Dealers Association reckons all the additional technology needed to achieve the 2025 goal will add $5,000 to the price of a new vehicle. Take somewhere in between—say, the $3,400 premium that

Camry hybrid owners pay—and any lifetime savings through lower running costs look even more difficult to recoup.

Much, of course, depends on what fuel prices do over the ensuing years. And that includes the price of coal and natural gas as well as oil. Remember, most of the new electric vehicles and plug-in hybrids that carmakers will have to build to meet the new CAFE requirements will be recharged overnight using electricity from power stations that mostly burn dirty coal or (if people are lucky) cleaner natural gas or uranium oxide.

The good news is that the bonanza in domestic discoveries of natural gas should help keep the lid on fossil-fuel prices for years to come. And if fuel prices come down in real terms, that will hurt sales of fuel-efficient vehicles.

Indeed, the way the NHTSA and the EPA have skewed the relief manufacturers get on vehicles of different sizes will pretty well ensure that American motorists will, once again, opt for trucks and SUVs rather than Prius-like vehicles. The deal done to buy off the industry's opposition to the new CAFE requirements granted manufacturers special credits for certain fuel-saving technologies. Meanwhile, pick-up trucks and larger SUVs were given what effectively amounts to a free pass.

The degree to which a vehicle has to comply with the NHTSA's fuel-efficiency and the EPA's carbon-emission requirements depends on the vehicle's "footprint" on the road. The larger the area (measured by multiplying the vehicle's track by its wheel-base), the greater the relief.

This was the fix that got Detroit's "big three"—which still rely on trucks and SUVs for the bulk of their profits—on board. No wonder Volkswagen and Mercedes-Benz have complained bitterly about the deal, while General Motors, Ford and Chrysler have now come out in support of the new CAFE proposal. An added sweetener for American manufacturers and trade unions was to exclude credits for the extremely frugal diesel engines that European carmakers have spent billions perfecting.

All told, then, will motorists across the country be getting 50mpg or more by 2025? Hardly. For a start, there is a world of difference between the testing methods used by the NHTSA for CAFE purposes and the EPA for determining a vehicle's fuel-economy figures for the window sticker that new vehicles must display in dealers' showrooms. Because of the difference, consumers will be paying for technology needed to achieve a CAFE average of 54.5mpg, but will be buying vehicles that achieve a real-world average of around 36mpg.

That assumes the CAFE target will still be in place by then. Because there is a limit to the number of years ahead for which the NHTSA can set fuel-economy standards, the new CAFE target will have to be reviewed by all parties in 2019. By then, much could have changed—in terms of energy price and availability, as well as engineering. Though a long shot, a breakthrough in storage technology would alter everything.

And by then there will, of course, be a new Administration in the White House with, quite possibly, an entirely different agenda concerning energy and the environment. As much as anything, insiders reckon the possibility of being let off the hook by a new Administration was the clincher that persuaded the motor industry to go along for the ride.

References

Allport, G. W., & Postman, L. J. (1947). *The psychology of rumors*. New York, NY: Holt, Rinehart and Winston.

Bayus, B. L., Kang, W., & Agarwal, R. (2007). Creating growth in new markets: A simultaneous model of firm entry and price. *Journal of Product Innovation Management, 24*(2), 139–155.

Gladwell, M. (2013). *The tipping point: How little things can make a big difference*. Boston, Little, Brown Group.

Rogers, E. (1962). *Diffusion of innovation*. New York, NY: Free Press.

Ryan, B., & Gross, N. (1943). The diffusion of hybrid seed corn in two Iowa communities. *Rural Sociology, 8*, 15–24.

The Myth of the Pioneer 6

> *I mean, when you're a pioneer and you are at the forefront of an offensive, you're going to be the most optimistic person.*
> Carlos Goshn, CEO Renault

Abstract
Most of today's market leaders entered the market 10–15 years after the pioneer. This is contradictory to schoolbook presentations where the advantages of pioneers are overstressed. Most market leaders go for big markets, make innovations in small steps and build up barriers through brands, patents, copyrights and business architecture.

6.1 Biased Statistics

The advantage of the market pioneer is well established in management literature. Several big studies provide statistical evidence for this hypothesis: For example, the PIMS (profit impact of market strategies) database discovered that mean market shares over a large cross-section of businesses are around 30 % for market pioneers, 19 % for early followers and 13 % for late entrants. PIMS data also show that 70 % of current market leaders are market pioneers. Similar estimates were derived from the database Assessor that included data from 129 consumer goods in 34 product categories. An Advertising Age Study supports these results: Out of 25 market leaders in 1923, 19 of them were still market leaders 60 years later. But Tellis and Golder questioned the hypothesis of the pioneering advantage in 1996 (Tellis & Golder, 1996). According to them, empirical evidence of the pioneer advantage has three limitations, all based on statistical bias:

1. Bias 1: All studies only collected data from surviving firms. But there are many pioneers that failed. Failures are important and change the statistics. These do not show up in the surveys. The failure rate of pioneers is close to 50 % both before and after World War II!
2. Bias 2: Most data has been collected from employees of the surviving firms by way of questionnaires. Careful analysis shows that people like to be perceived as pioneers. Their answers are biased.

3. Bias 3: The definition of the pioneer has been too soft in most studies defining the concept of pioneers "as the first to enter the market", based on a personal theory at best, targeted to a specific effect at worst.

In the past, the label pioneer was used for dominant firms that pioneered a new concept or a new market segment. Tellis and Golder narrow the definition down to "first to sell in a certain product category", where a second supplier follows with a competing brand and a close substitute. Examples of such product categories are microwaves, copiers or disposable diapers. This definition of the pioneer clearly differentiates the time order of market entry.

The myth of the pioneer is a typical example of the difficulties in obtaining precise data from questioning people in organizations. The data are not based on facts, but opinions, which are subjective and built on personal theories. A vivid example of biased samples are the British statistics on German air defence cannons and their impact on allied bombers: According to statistics the plane's centre was never hit. Not shown by the statistics is the explanation that all planes that were hit in the centre crashed because their kerosene tanks were located at the centre. Destroyed planes of course did not show up in the statistics.

Historical analysis of pioneers, or better of "first sellers in a category" based on published catalogues not on interviews, shows that market pioneers have an average market share of 10 % and only 11 % of them are later market leaders (Golder, 1993). Examples of pioneers that were the first in a product category are:

- Howard Deering Johnson (1897–1972), an American hotelier who was the pioneer of the restaurant-franchising concept in 1935: An operator used name, food and logo in exchange for a fee. The "HoJo" chain restaurants expanded rapidly.
- Disposable diapers only penetrated the market around 1950, but were already introduced by the Swedish company Pauliström Bruk in the 1930s. First, they were only available as a luxury product. Then, two mothers who are famous today being the pioneers of disposable diapers, Marion Donovan in 1946 in the US and Valerie Hunter Gordon in 1947 in the UK, perfected the system. In 1949, the company Johnson & Johnson entered the market with the first disposable diaper called "Chux".[1] Today most people believe that today's market leader, Procter and Gamble, was the pioneer. Procter and Gamble had a market share of more than 30 % in 2009 with its brand Pampers.
- Micro Instrumentation and Telemetry Systems (MITS), an American electronics company founded by Ed Roberts and Forest Mims, began manufacturing personal computers in 1975. Roberts developed the first commercially successful home computer, the Altair 8800. Paul Allen and Bill Gates then started writing the software for the Altair computer.

Nobody remembers Howard Deering Johnson, Marion Donovan or Ed Roberts who were very early pioneers in their fields. The enduring market leadership of their products is not connected to their names.

---

[1] NZZ, April 29th, 2007: Die Wegwerfwindel am Wickeltisch ohne Konkurrenz.

6.2 The Early Adopters

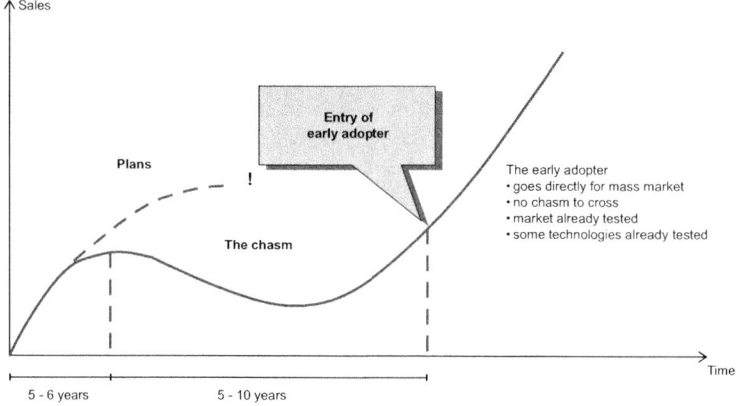

Fig. 6.1 The early adopter: challenge of timing and vision of mass market

6.2 The Early Adopters

Far more successful than pioneers are early adopters (Fig. 6.1). They have a much lower failure rate, their market share is three times higher than those of the pioneers, but they enter the market on average 13 years after the pioneers. And then – nobody remembers the pioneers anymore! It is nice to be a pioneer, but in many cases it does not pay off!

Early adopters have some distinctive advantages compared to pioneers:
- They have a vision of the mass market. They see what the new technology can change through the eyes of the customer who already uses the new technology and can give real feedback on the product. A pioneer instead has to rely on potential estimates of surveyed future customers.
- They are persistent. Early adopters understand how important it is to cut costs from the beginning. For example, it took Sony 20 years of R&D to lower the cost of a video recorder from USD 50,000 to USD 500.
- They know how much it is going to cost to introduce a new product to the market.
- They push innovation in small steps as soon as they have more information about the real use of the new product on the market.
- They understand how to leverage their assets and their core competencies. Brand, distribution networks and existing service organizations may help to penetrate the market much quicker than a beginner is able to do.

Many early adaptors follow a trickle-up strategy (Fig. 6.2): They go immediately for the big volume and do not try to milk a niche at the beginning. Toyota went into the mass market of hybrid engines with its Prius. Several years later they started in

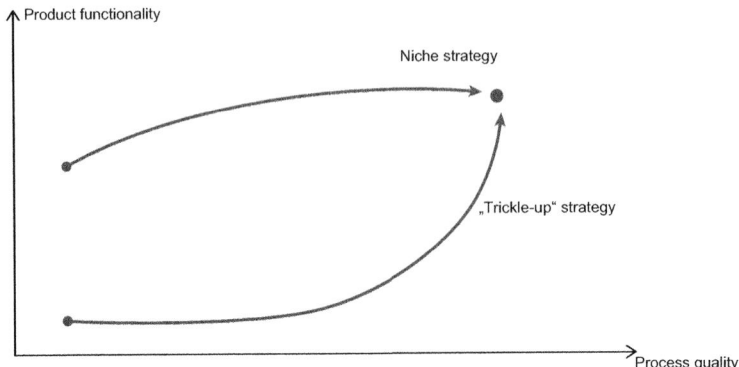

Fig. 6.2 Established successful enterprises are producing quality, but have to lower the costs of a high-quality product

the top niche with Lexus. It is much simpler to upgrade a cheap product than to lower the cost of an expensive high-quality product.

Many traditional enterprises are known for their high product functionality right from the beginning. They enter the price-functionality-pyramid from the top. To achieve low cost they have to improve process quality, sometimes at the expense of product functionality. Toyota's Prius started with low product functionality, low cost and top process quality.

Early adopters, such as Toyota, know how important it is to go for big volume immediately. They improve their products in small steps as soon as they have more information from the market and create entry barriers through low cost for potential competitors. They even subsidize their market entry to make it clear for competitors that it is hard to make money in the new market.

Some companies develop new technologies, but do not see the market opportunities. They do not enter the mass market at all. XEROX failed to succeed with its office computer business: Besides developing many modern computing technologies, such as the graphical user interface (GUI) or laser printing, XEROX also invented the Xerox Alto in 1973, a small minicomputer similar to a modern workstation or personal computer, our PC. This machine can be considered as the first true PC, given its combination of a cathode-ray-type screen, mouse-type pointing device, and a QWERTY keyboard. But the Alto was never commercially sold, as XEROX itself could not see the sales potential of it. It was, however, installed in Xerox's own offices worldwide and the offices of the US Government and military. In 1979, Steve Jobs made a deal with XEROX's venture capital division: He would let them invest USD 1 million in exchange for a look at the technology they were working on. Jobs and the others saw the commercial potential of the WIMP (Window, Icon, Menu, and Pointing device) system and redirected development of the Apple Lisa to incorporate these technologies. Jobs is quoted as saying, "They just had no idea what they had." In 1980, Jobs invited several key

6.2 The Early Adopters

Table 6.1 Innovator – imitator: success and flops (Teece, 1986)

	Innovator (pioneer)	Imitator (follower)
Success	Pilkington (float glass)	IBM (PC)
	DuPont (Teflon)	SEIKO (quartz watch)
Flop	XEROX (PC)	DEC (PC)
	RC Cola (diet cola)	Kodak (digital photos)

XEROX researchers to join his company to fully develop and implement their ideas.

RC Cola (Royal Crown Cola) was the first beverage company to introduce coke in a can in 1954 and later the first company to sell a soft drink in aluminium cans. The company introduced the first diet coke in 1958. Sales were disappointing largely due to the inability of the RC bottling network to secure distribution for the product in single-drink channels. It was quickly discontinued with the exception of Australia, New Zealand and France. Today it is only available in New Zealand and parts of Australia. Coca Cola and Pepsi followed and imitated the innovation immediately and profited from the pioneer's preparatory effort (Table 6.1).

A very successful pioneer is the Pilkington Group Limited, a multinational glass manufacturing company known for the invention of Sir Alastair Pilkington's float glass process between 1953 and 1957. The float glass process was a revolutionary method of high quality flat glass production. It floats molten glass over a bath of molten tin, avoiding the costly need to grind and polish plate glass to get flat pieces. Pilkington was almost killed by his invention: It cost far more than expected. Pilkington then allowed the float process to be used under license by numerous manufacturers around the world.

An invention that was created by accident and then became a successful innovation was Teflon. Kinetic Chemicals, a joint venture between DuPont and General Engines, patented it in 1941 and registered it as a trademark in 1945. In 1948 DuPont was producing 900 million tons of Teflon. In 1954 the first kitchen pan was coated with Teflon under the brand name Tefla. Today DuPont describes itself as a global science company that is still known for its innovation of Teflon.

Perhaps the best-known example of a follower is IBM's PC. In its early years IBM intended to follow a strategy of rapid imitation, transferring the most important technical knowledge from universities and competitors. Although IBM's first data-processing computer, model 702, was troublesome and unreliable, an improved version, the model 705, was soon introduced in 1954. By 1955 IBM was selling more computers than UNIVAC, the company that had built the first business computer 4 years ahead of IBM.

Another successful follower is SEIKO, a Japanese watch company. Even though the world's first prototype of analogue quartz wristwatches was revealed in 1967 in Neuchâtel Switzerland, Seiko followed and produced the world's first commercial quartz wristwatch, the Astron in 1969. SEIKO is still a successful producer of digital watches.

Compared to IBM, PCs of Digital Equipment Corporation (DEC) who tried to follow as well are relatively unknown today. In 1998 DEC was acquired by

Compaq, which merged with Hewlett-Packard in 2002. Other parts of DEC were sold to Intel.

Thus reality is not black and white, but grey. You can be successful or produce a flop whether you are a pioneer or an early follower. Once again, statistics and averages do not tell us anything about a specific company.

6.3 Benefits from Competitor's Homemade Barriers

Early adopters benefit often from homemade entry barriers of their competitors. Market leaders are often complacent and rest on their laurels. Thus many of their innovation barriers are self-made:
- IBM had second thoughts about introducing PCs because they cannibalized their mainframes. IBM feared losing its existing business.
- Some market leaders rest on their laurels and slow down innovation. This is particularly true for leaders close to a monopoly. The leaders in cement production have been in an oligopoly for many years and have outsourced R&D to a large extent.
- Some big companies build up unnecessarily sophisticated approval processes and lose speed due to bureaucracy, as evidenced by some car manufacturers and suppliers.

Gillette's history shows how companies can overcome these barriers. Since 1903, men used straight or expensive safety razors. After 1903, King Gillette introduced a safety razor with disposable blades. The Gillette Company marketed its razors and blades successfully by targeting the US government to buy them for troops in the two world wars. Thanks to patents for their initial products, Gillette dominated the mass market for half a century but was shaken when Wilkinson entered the market in 1962 with a steel blade that lasted three times longer. Due to the lack of financial resources, Wilkinson Sword could not destroy Gillette. But Gillette, now galvanized by its Wilkinson experience, even innovated at the cost of cannibalizing its own established products. Other competitors entered the market but Gillette kept on continuously innovating with twin blades and so on. At any given time Gillette has at least 20 shaving products in the pipeline. Each day about 200 employees personally test new shaving technologies. This emphasis on testing together with its innovation speed probably supports Gillette's dominance in the market. Gillette is one example of small, incremental innovations in design, manufacturing, and marketing over many years. Much too often new products are described as "breakthrough innovations" when all they really are just small modifications of existing products. But in reality, many market leaders achieved their dominance through small and incremental innovations.

6.4 Switching Costs

Beside its continuous innovations, Gillette creates another entry barrier: When consumers have bought the Gillette razor, disposable blades prevent customers from buying another razor, as it would induce switching costs. Thus Gillette can offer their razors below cost, whereas the blades are sold well above costs. Printers from Hewlett Packard work the same way: Once bought, customers have to buy HP cartridges. It is a captive business!

Creating entry barriers using switching costs is typical for captive markets. In a captive market, the potential consumers face only few competitive suppliers. Captive markets result in higher prices and less diversity for consumers. The term therefore applies to any market where there is a monopoly or oligopoly. If a company is able to create high switching costs with an increased level of customer dependency, the company can shape the market.

6.5 Entry Barriers

If there is ideal competition in a market, entry barriers do not exist. As this is not the case in any real world, barriers to entry are far more important than usually perceived. Generally, three types of entry barriers exist:
- Laws: e.g. Intellectual property rights
- Reputation: Brands
- Organizational architecture: Business models

Barriers can be built up with a proprietary standard, through patents, through qualified personnel, through speed in innovation or above all through tacit know-how. Even after decades of car production in China, Chinese producers still have big problems building "Chinese" cars. Sometimes it is better to be dependent on key people than to have all the know-how well documented and distributed among all the employees. Pharmaceutical companies, such as Novartis, are typical examples of companies which create entry barriers not only with patents but with qualified knowledge workers as well.

The most effective barriers exist in mature industries: If you aim to compete with Lafarge or Holcim in cement manufacturing you have to spend at least USD 100 million that is what a cement factory costs. You need a quarry and of course a license to use the quarry! You have to build up the reputation these companies have and do not forget that the existing companies are able to retaliate for a long period. Many mature industries have strategic assets.

A consistent organizational architecture, a specific business model, can also prevent others from copying a business easily. Zara, the Spanish clothing and accessories retailer, has built up a unique business model: Compared to the 6-month industry average, Zara just needs 2 weeks to develop a new product and get it to the stores. It launches around 10,000 new designs each year. Zara has resisted the industry-wide trend towards transferring fast fashion production to low-cost countries. Perhaps its most unusual strategy was its policy of zero

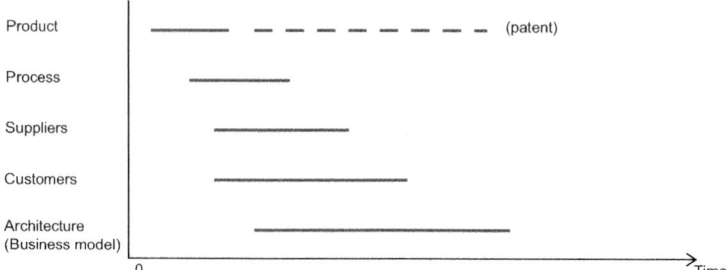

Fig. 6.3 Sustainability of innovation barriers

advertising; the company preferred to invest in new stores in prime locations instead. This has contributed to the idea of Zara as a "fashion imitator" company and low cost, not low price, products. Zara was described by Louis Vuitton Fashion Director Daniel Piette as "possibly the most innovative and devastating retailer in the world." and as a "Spanish success story" by CNN.

Technical advantages in products, processes, suppliers or customers can be best sustained when combined with organizational innovation or architecture. Successful protection is always a combination of different concepts (Fig. 6.3).

In mass markets there is only little differentiation, but fierce competition, low prices and in many cases low profits, at least per product. In contrast, products in new markets have specific, limited product functionality, high prices and few applications. At this time in the life cycle the new product needs a visionary who believes in the product and finds a way to the mass market. Only in mass markets, can economies of scale be exploited and experience gained in order to overcome low process quality, the key to product cost reduction. Only in a mass market can the full potential of a new product be exploited. This is how VHS drove the AMPEX video recorders out of the market.

The Economist: Samsung: The next big bet

The world's biggest information-technology firm is diving into green technology and the health business. It should take care; its rivals should take notice©
The Economist Newspaper Limited, London (Oct 1st 2011)

IN 2000 Samsung started making batteries for digital gadgets. Ten years later it sold more of them than any other company in the world. In 2001 it threw resources into flat-panel televisions. Within four years it was the market leader. In 2002 the firm bet heavily on "flash" memory. The technology it delivered made the iPhone and iPad a reality, and made Samsung Apple's biggest supplier—and now its biggest hardware competitor.

The handsome payoffs from these ballsy bets made the South Korean company a colossus; last year its sales passed $135 billion. Now it is embarking on a similarly audacious plan to move away from electronics into technologies where it barely has a presence today. It intends to spend $20 billion over ten years on solar panels, light-emitting diodes (LEDs) used for lighting, electric-vehicle batteries, medical devices and biotech drugs. These businesses shift Samsung away from easily substitutable gadgets towards more essential industrial goods (see table)—or from "infotainment" to "lifecare", as the company puts it. Just as electronics defined swathes of the 20th century, the company believes green technology and health care will be central to the 21st.

With these plans Samsung sees itself bringing technologies that are vital for society into much broader use. The company has always had an eye for more than just the bottom line, seeking both to epitomise and to further the progress of its home country. Now it talks idealistically of improving the world by driving down the costs of zero-carbon power and providing poor countries and rural areas with medical equipment and drugs that they cannot afford today.

Fresh fields
Samsung's new business areas

Sector	Investment, $bn	Ownership	Targets for 2020 Sales, $bn	Jobs	Status
Solar panels	5.1	100% Samsung SDI	8.5	10,000	Production began in January
LED lighting	7.3	50% Samsung Electronics, 50% Samsung Electro-Mechanics	15.2	17,000	Already selling in South Korea
E-vehicle batteries	4.6	50% Samsung SDI, 50% Bosch	8.7	7,600	Initial operations began in November 2010
Biotech drugs	1.8	40% Samsung Electronics, 40% Samsung Everland, 10% Samsung C&T, 10% Quintiles	1.5	1,000	Factory to begin in 2013; developing biosimilars now for patents expiring in 2016
Medical devices	1.0	100% Samsung Electronics	8.5	10,300	Blood-testing unit available, X-ray machine ready in 1-2 years, acquired ultrasound maker

Sources: Samsung; *The Economist*

But the plans are also an ambitious industrial power play, one that challenges some of the world's biggest companies. Success would raise Samsung to new heights. Failure could lead to the firm losing what it already has, no longer able to flourish just as a maker of commodity gadgets and components.

The 83 firms that are tied together in Samsung's remarkably complex structure provide 13% of South Korea's gross exports. Samsung Electronics, the biggest of them, started making transistor radios in 1969, and has since evolved into the world's leading manufacturer of televisions and much else. It is on track to unseat Nokia as the biggest maker of mobile phones by volume next year. Interbrand, a consultancy which seeks to calculate brand value, puts it in the world's top 20, ahead of Sony and Nike. It has come second only to IBM in the number of patents earned in America for five years running.

Yet Samsung wants to diversify away from consumer electronics, a market that suffers from falling prices, thin margins, fast product cycles and fickle customers. Chinese rivals may do to Samsung what Samsung did to Western and Japanese firms in the past. "The majority of our products today will be gone in ten years," Samsung's patriarch and chairman, Lee Kun-hee, told executives in deliberately alarmist tones last January.

To survive, he said, the company must not only go into the new businesses it has identified, but open itself up to work with partners and even make acquisitions. Samsung has long been a closed world from that point of view, a disposition reinforced after the disastrous acquisition of a PC maker in the 1990s. But now the company knows it needs new skills, sales channels and customers.

Doing it the Samsung way

By 2020 Samsung's Mr Lee wants the five new business areas to provide $50 billion of revenue, and Samsung Electronics to be a $400 billion company (for all his provocations to his staff, there are still going to be a lot of flat screens and memory sold). It is a brash goal, admits Inkuk Hahn of Samsung's strategy team. But ten years ago people were incredulous when Mr Lee insisted that Samsung, which then had sales of $23 billion, could be the number-one technology company, with sales of $100 billion. It claimed that crown just eight years later. "This is why you have to believe us," Mr Hahn insists.

The new businesses look remarkably disparate, but they share a need for big capital investments and the capacity to scale manufacture up very quickly, talents the company has exploited methodically in the past.

Samsung's successes come from spotting areas that are small but growing fast. Ideally the area should also be capital-intensive, making it harder for rivals to keep up. Samsung tiptoes into the technology to get familiar with it, then waits for its moment. It was when liquid-crystal displays grew to 40 inches in 2001 that Samsung took the dive and turned them into televisions. In flash memory, Samsung piled in when new technology made it possible to put a whole gigabyte on a chip.

When it pounces, the company floods the sector with cash. Moving into very high volume production as fast as possible not only gives it a price advantage over established firms, but also makes it a key customer for equipment makers. Those relationships help it stay on the leading edge from then on.

The strategy is shrewd. By buying technology rather than building it, Samsung assumes execution risk not innovation risk. It wins as a "fast follower", slipstreaming in the wake of pioneers at a much larger scale of production. The heavy investment has in the past played to its ability to tap cheap financing from a banking sector that is friendly to big companies, thanks to implicit government guarantees much complained about by rivals elsewhere.

From crisis to crisis

Competitors also balk at the way that Samsung scales up quickly to supply parts to other firms as well as to price its own gadgets keenly. Supplying the rest of industry drives down Samsung's costs yet further, with its rivals in effect financing its success. This strategy can create problems. Samsung is Apple's most important supplier in the smartphone and tablet-computer markets. Samsung components, which include all the product's application processors, account for 16% of the value of an iPhone. It is also Apple's greatest competitor in those markets. Apple is now suing the socks off the company for copying the look and feel of its products. At the same time it is urgently seeking new ways to diversify its supply chain.

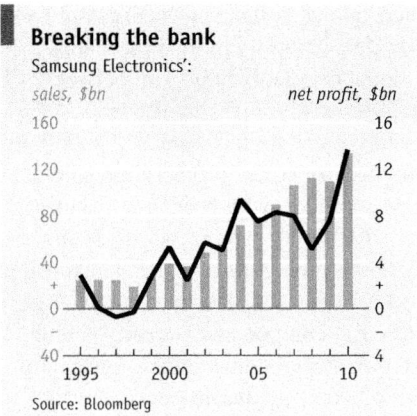

Breaking the bank
Samsung Electronics':
sales, $bn / net profit, $bn
Source: Bloomberg

Many companies saw the potential of technologies such as liquid-crystal panels, flash memory and rechargeable batteries. But few could or would invest billions in a single shot. That Samsung could is in large part due to a cult of personality around Mr Lee, who likes to keep things shaken up. "Change everything but your wife and children," he exhorted managers in 1993. Three years later he lit a bonfire of 150,000 gadgets because some were defective. Other bosses often need to face a crisis—a "burning platform", in the memorable phrase of Stephen Elop, Nokia's boss—before they make changes. Samsung does so when things are going well. The company has pushed out older managers and restructured its divisions over the past two years despite posting record profits even in the global financial crisis.

Management by perpetual crisis is perhaps a reflection of the company's national roots. In 1960, when the Samsung companies were taking off, South Korea, battered by recent war, had a GDP the same size as Sudan's; its last dictatorship fell only two years before the Berlin Wall. Today, though it enjoys one of the world's highest living standards, South Korea is still an emerging market in some ways, with endemic corruption and some economic structures that border on the feudal.

Samsung, like its host country, has a foot in both the industrialised and developing worlds, which it has used to its advantage. While it has always produced things for major IT firms and Western consumers, it has aimed products at poor countries, too. This not only gave Samsung scale, but also market shares in the world's fastest-growing economies. Whereas Western firms reeled in the recent recession, Samsung flourished, buoyed by sales in markets that never stopped growing.

From laptop to rooftop

Some of the five new businesses Samsung has set its sights on are not that far from what the company does already. Its experience in semiconductors and flat-screen televisions fits easily with solar cells and LED lighting: the technology, materials and production processes are similar. Likewise, its expertise in batteries for gadgets smooths the way for making car-sized ones. The firm wants to apply the magic of ever cheaper chips to medical devices as it did mobile phones. Even drugs aren't so far afield when one sees them in business-process terms: high-volume manufacturing with low defect rates. In all these fields Samsung believes it can sit—rather as Korea does geographically—in between China, with its cheap products, and Japan, with its costly, high-quality ones.

In solar energy Samsung plans to make panels for both domestic and industrial use. Producing panels for "utility-scale" projects may allow it to lower prices for the residential market. Changsik Choi, who heads the business, also speaks optimistically of a "brand halo effect": consumers whose living rooms are stuffed with Samsung products may choose the company for their rooftops too.

Samsung's dominance of the television market has already made it the world's second-largest maker of LED components (Japan's Nichia is the first). Since they consume a fraction of the power of conventional light bulbs, last longer and avoid some of the drawbacks of compact fluorescents, the first-generation alternative, LEDs are expected eventually to become the norm for all sorts of lighting; the market is growing by 65% a year. Samsung already sells LED lighting in South Korea and plans soon to expand abroad. In this market it will hew to its strategy of supplying parts to others, thereby lowering costs for its own products.

In electric-vehicle batteries Samsung has joined forces with Bosch, the world's biggest supplier of car parts and a fount of expertise on power- and engine-management. Samsung sees their partnership, SB LiMotive, as crucial since the car business relies on close ties between carmakers and their suppliers. Some carmakers, like Nissan and Toyota, will continue developing their own batteries, but Samsung thinks that many carmakers will not want to be in the battery business, just as they are not in the petrol business, and that they will be a rich source of demand. Chrysler and BMW are among SB LiMotive's first customers.

For medical devices Samsung aims to use information technology to lower costs, add features and make devices accessible to more people, particularly the poor. For example, it is developing X-ray machines that expose patients to less radiation and do away with physical film. Last year Samsung began selling a machine for testing patients' blood chemistry that is smaller, cheaper, uses less power and offers more functions than rivals' devices. In April it bought Medison, a South Korean maker of ultrasound equipment, as a way to get further into the market: it is looking at buying body-scanner firms too.

In biotech drugs the company plans to begin as a contract manufacturer of biosimilars (generic versions of biotech drugs) and has partnered with Quintiles, a drug outsourcer. The strategy lets Samsung gain experience while assuming little commercial risk. It is building a factory outside Seoul and has already begun developing biosimilars for medicines with patents that expire in 2016.

Incumbents and incomers

The markets are certainly promising, but they entail huge risks. Nor is the size of Samsung's commitment quite on a par with the overwhelming force it has deployed in the past. The solar and LED businesses already struggle with oversupply, meaning Samsung may get walloped by the same dramatic price erosion as it has seen in liquid-crystal flat panels. Electric-vehicle batteries may be in similar straits if demand for the cars they might power remains sluggish. They are also in the crosshairs of Chinese companies, as are medical devices and drugs. In a bid to escape the vagaries of consumer electronics, Samsung may be ploughing headlong into the areas most ripe for invasion by a new breed of emerging-market titans.

Acquisitions, a way of life in the drug business, are also a challenge: knowing what to buy and when is a skill that Samsung has never developed. The same applies to dealing with government regulators: Samsung's towering importance at home may give it a false confidence in its ability to handle governments elsewhere.

Its position as a domestic titan could be a hindrance in other ways. Working with partners entails sharing information and a view of joint success that is at odds with its insular corporate culture. The international talent the company will need to attract is also less likely to be moved by the admonishments and appeals to national grandeur that Mr Lee has used to build Samsung's success. They might, indeed, find such things wearisome.

Samsung's rivals are ready for a fight. Philips and GE have been preparing to compete with firms in emerging markets for years, devising cheap products and building on existing relationships with clients. Toshiba plans to spend an extra $9 billion in the energy and environment sectors over the next three years on top of its normal capital expenditure, research and acquisitions. Fumio Ohtsubo, Panasonic's boss, praises his Samsung rivals for their low prices but believes his company develops superior technology. "If we can get the same conditions in terms of free-trade agreements, low corporate taxes and other incentives, then we should be able to compete," he says.

In medical devices Samsung will be up against firms like Philips, Siemens, Toshiba, Hitachi and GE (for which Samsung made medical equipment between 1984 and 2004). GE's Indian office famously reduced the cost of an electrocardiogram machine from $2,000 to $400. And the fact that hospitals prefer to buy different equipment from a single vendor so that, in principle, everything works together puts a maker of this-and-that at a disadvantage, even if it is cheap.

Perhaps the biggest challenge, though, will be one of succession. The 69-year-old chairman's son, Jay Y. Lee, 43, was named president last December. Educated in Japan (like his father and grandfather, the firm's founder) and at Harvard Business School, he has been groomed from the start. His first test will be reforming the jumble of opaque, interlocking relationships and conflicts of interest that passes for Samsung's corporate governance.

The "Samsung group", as it is often known, has no legal identity. The 83 firms sit under an umbrella company called Everland, in which the Lee family has a controlling 46% stake. The family also has minority positions in other Samsung firms, which often hold shares in other members of the group, and indeed in Everland. For example, the family and related interests own 21% of Samsung's life-insurance firm, which owns 26% of its credit-card business, which in turn owns 26% of Everland. Get it? Nobody other than the Lees really does.

The company must change if only to avoid South Korea's devastating 50% inheritance tax after the elder Lee's passing (his father died at the age of 77). That would further whittle the family's stakes, notes Shaun Cochran of CLSA, a broker. He expects a holding company to be formed, so investors have clearer exposure to the different parts of Samsung's businesses. The younger Lee will also need to root out corruption, which his father often complained about without rising above it; the elder Lee's 2008 conviction for tax evasion was pardoned in 2009 on the ground of his importance to the country.

When the dealing's done

Chairman Lee's fear is that successful companies get flabby when they hit middle age. He saw that in Sony, founded in 1946, which has been struggling since the 1990s. Samsung Electronics turned 40 in 2009, which prompted Mr Lee to lay the groundwork for the five new growth areas. Diversification is essential. In the mid-1990s almost all of its profits came from DRAM memory chips: when the market soured in 1996, its profits shrivelled by 95%.

Samsung may be swapping "infotainment" for "lifecare"—but it is still in the hardware business, and that may leave it more vulnerable than it thinks. Many of today's computer and electronics giants are getting out of the manufacturing businesses altogether. IBM has shifted to services, trailed by Japan's Fujitsu, while Philips and Siemens both sold their IT businesses to focus on other areas. But getting out of things is not something Samsung is good at. Despite a commitment to perpetual crisis, a mixture of implicitly subsidised capital, weak shareholder pressure and family control has allowed it to stick too long with dodgy decisions—such as its move into cars, brought short only by the Asian financial crisis, and its only-now-ended commitment to hard-drive manufacture.

Even with a $20 billion bankroll, bets can be spread too thin. Perhaps the biggest risk for Samsung is not that none of its wagers will win, but that it won't be able to stop betting on the ones that don't. Knowing the right time to bet is a great gift. So is knowing the right time to walk away.

References

Golder, P. N. (1993). Pioneer advantage: Marketing logic or marketing legend? *Journal of Marketing Research, 30*, 158–170.
Teece, D. J. (1986). Profiting from technological innovation: Implications for integration, collaboration, licensing and public policy. *Research policy, 15*(6), 285–305.
Tellis, G. J., & Golder, P. N. (1996). First to market, first to fail? Real causes of enduring market leadership. *MIT Sloan Management Review, 37*(2), 65–75.

Pirates, Pioneers, Innovators and Imitators 7

> *It's more fun to be a pirate than to join the navy.*
> Steve Jobs

Abstract
Technology opens up new opportunities for good and for bad. At the beginning of a lifecycle only freaks are interested, nobody cares. Later pirates commercialize, make money, societal impacts grow, some big companies cry foul. Rules, laws, dominant designs and standards set in, they provide predictability, the big money can move in, transaction costs go down; technology matures, until a new wave of technology starts the game again.

7.1 The Technology Life Cycle

Douglas C. North, Nobel Prize Winner in economics 1993, remarked: "Undergirding ... markets secure property rights, which entail a political and judicial system to permit low cost contracting" (North, 1991). He stressed the importance of generally accepted rules for economy. Without rules transaction costs go up and economy loses efficiency. Nevertheless, it takes time to develop these rules, sometimes hundreds of years. New technologies may contain new inadequacies, allow new types of crimes and thus trigger new laws, or simply need new rules to ensure low- cost contracting among market members. For example, mobile phones can be used to trigger bombs but also to trace terrorists or drugs and therefore need to be monitored by a central authority to ensure safety and security.

Without rules a sustainable mass market would not be possible. Rules that tame technologies usually develop in four phases:

1. First there is innovation. Nobody cares too much about rules, neither the innovator nor the society. The impact at the beginning is simply too unimportant. New markets and new products start small. Nobody makes big profits. Companies need seed money. In 1837, for example, Samuel Morse showed the telegraph to the American congress, and nobody was interested.
2. After some years – it may take 5–40 – somebody becomes aware of the commercial potential of the new technology and starts to make money

(commercialization). Nobody is worried; the business is still too small. Possibly some venture capitalists start looking at these new opportunities and some competitors may already take measures. Sir Francis Drake, an English sea captain, privateer, navigator, slaver, and politician of the Elizabethan era, saw the opportunity to make profit by pirating Spanish merchant ships. He brought back spices and tea worth billions, measured in today's terms, making him a hero to the English but a pirate to the Spaniards who called him El Draque.

3. In the third phase, after another 2–10 years, market and societal impacts grow. More and more entrepreneurs see the potential and start new businesses, as in the dot-com bubble around 2000. The newcomers start hurting established companies. Functionality is prime at this moment and with many competing solutions, the market grows as does the impact on society. Due to widespread creative anarchy, nobody wants to invest big money: It is too risky. The question is still whether there will be a standard or a dominant design? The community starts asking for rules. The general understanding of the new technology is advanced enough to allow the development of laws, standards and good practices. Monsanto, a leading producer of genetically engineered seed, would be glad to have worldwide accepted standards in genetically modified food.

4. In the fourth phase, rules, laws, standards and dominant designs tame the technology. The community knows how to behave. Without rules, no sustainable mass market would be established as companies demand rules and security to be able to invest. A dominant business model develops and is protected by accepted rules until a new technology cycle starts again.

In summary, rules develop over the technology life cycle: When a technology emerges, pioneers and pirates do not perceive a gap between law and technology. The more the technology is commercialized and the more anarchy prevails, the bigger the gap between law and technology. Rules close the gap and enable the emergence of big enterprises (Fig. 7.1).

Pioneers and pirates take over technologies from innovators. But what is the difference between pioneers and pirates? Whereas pirates take advantage of the gap between technologies and laws, pioneers take advantage of their exclusive capability to see and realize the potential of the new technology. Pirates need specific customers to be successful: These customers take the technology as it is, solve a problem and do not ask too many questions about side-effects and regulations. Most importantly, customers should not ask for improvements. These prerequisites are mainly given if customers are under time pressure to move.

7.1.1 Pirates and Pioneers in Ship Technology

The development of ship technology is a good example for pirates and pioneers: It experienced a huge improvement between 1200 and 1500: Ships' planking developed from clippers with thick planks to caravels with frames and planks to ships that were lighter and bigger and could be managed with a smaller crew. After 1250, navigation made a big step forward: Accurate sea maps with nautical instructions,

7.1 The Technology Life Cycle

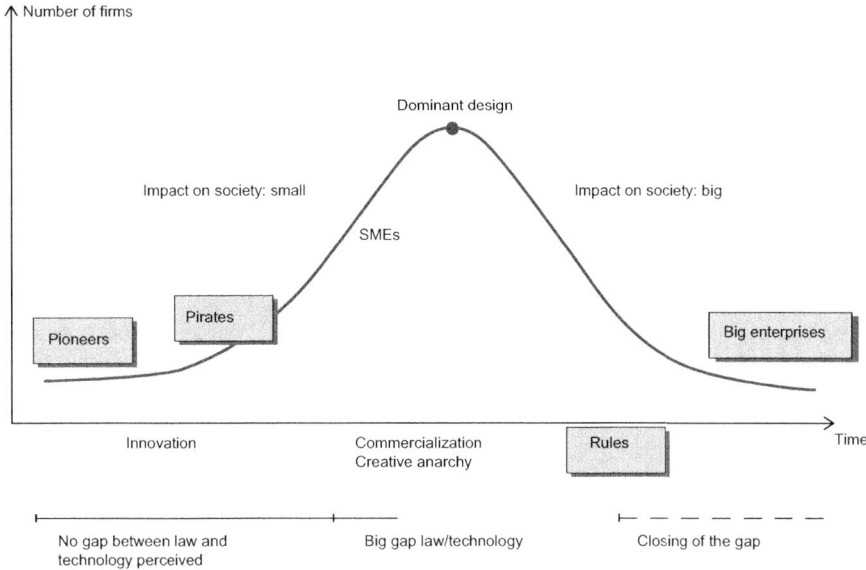

Fig. 7.1 Rules and regulations develop along the technology cycle

the Portulanos, were used. Even though the compass had been invented and used in ancient China since the eleventh century, it was not accepted in Europe until 1400. Later the quadrant was used to ascertain the position of the ships from measurements of the positions of the sun, moon and stars. The application of science and engineering as a complete system had improved ship technology in small steps and made it a success.

In Portugal, Henry the Navigator was the pioneer who pushed navigation and ship building to new frontiers. In his lifetime, he managed to transform his country of origin into a worldwide leading sea power. Around 1420, he started an R&D Centre in the South of Portugal for the furthering of European exploration and maritime trade with other continents. Between 1424 and 1456 his expeditions discovered Madeira, the Azores, Cape Verde Islands and Guinea. After his death in 1460 sailing prevailed as a rigorous art. Other pioneers followed Henry, such as Columbus from 1451 to 1506, Amigo Vespucci from 1454 to 1512 and Ferdinand Magellan from 1480 to 1521. Many of these pioneers started their expeditions, but only a few came back.

Cortez and Pizarro who looted the Americas used Henry's navigation and shipping technologies to make money. They were the first who saw the profit. Later, pirates such as Francis Drake (1540–1596) were appointed by their governments to control the seas in their own interest. At that time commercialization was defined as: "Define the territory and make your claim." Pirates moved without restrictions, but were also severely and unscrupulously punished.

It took some 200 years to set up worldwide regulations with the declaration of Paris in 1856 to put an end to piracy and to allow commercial trade to flourish. The modern sea rights were born, taking their final shape in 1958 in the Sea Rights Convention of Geneva. In the end, even pioneers and pirates strove for order: Once capitalists, owners of valuable assets, they looked for property rights and someone who would protect them. Usually this was the government. The more prosperous market members became, the more they asked for rules. The story about the pirates of the seas 500 years ago has repeated itself again and again. Rules increase predictability, even negative ones: Once nuclear power is regulated out of the market, companies can stop investing in that field of technology. The worst is an open situation where nobody knows whether the rules will change or not. A recent example is the music industry.

7.1.2 Pirates and Pioneers in the Music Industry

In 1877, with the invention of the phonograph by Thomas Edison and the radio, the way music was heard changed forever. Around 1900 the technology spread around the world and the "record industry" developed. In 1923 rules were set in the copyright law: Artists started to contract with big music companies by selling their rights for a lump sum to a distributor who promoted them and sold their music. A business model that was profitable for the "Big 6" – BMG, EMI, Polygram, Sony, Universal and Warner who dominated the industry at the end of the 1980s. In 1996 they controlled more than 75 % of the world music market. In mid-1998, PolyGram merged into Universal Music Group (formerly MCA), reducing the leaders to the "Big 5". They became the "Big 4" in 2004 when Sony acquired BMG. Musicians and actors were exploited by the big retailers, but were too weak to change the system. But some years later the tide turned and technology changed everything with amazing speed.
- Peer-to-peer (P2P) technology solved the problem of music storage and created huge increasing returns. Using P2P opens your files to the public for free and leads to cheap decentralized storage.
- MP3, a new standard, was developed at the Fraunhofer Institute in Germany. It reduced download times over the Internet by a factor of 10–20 with an intelligent compression algorithm.
- Bandwidth increased dramatically with glass fibres.
- PC literacy was acquired by billions of users.

Within 4 years after introduction of MP3, 26,000 illicit music sites opened and offered 200,000 songs for free on the Internet. The music industry became worried because some artists circumvented the "Big 6" and distributed their new creations directly online. Many venture capitalists already trumpeted the death of the traditional music industry. In fact, consumers spent less money on recorded music. Revenues dropped causing layoffs and forcing record companies to seek new business models. In the first two quarters of 2005 physical formats' sales, such as

CDs, decreased by −6 % while sales of digital downloading increased by some 300 %.

The traditional music industry fought back, first with lawyers only, then by accepting the new technology. But first, pirates made money during the years of anarchy.

One of them was Shawn Fanning, a drop out of North Western University who saw the opportunity. He started Napster in 1999 and within a few days he had approximately 5,000 customers. One year later he counted 40 million users and 20–30 % of the US university bandwidth was used to download music. The record industry took aggressive action against file sharing and succeeded in shutting down Napster in 2001. The big music companies charged Fanning with causing billions of dollars in losses. After some time of creative anarchy, rules had taken over again. However, this failed to slow the decline in revenues and some academic studies suggested that digital music downloads did not cause the decline (Saw & Koh, 2005, Cowen, 2008). Finally, musicians and Fanning himself had to accept that they had to solve the same problem as the "Big 6": How to charge the customer?

Today the rules are set, the standards as well. Fanning is currently working with Universal Music Group and Apple selling music with great success. Legal digital downloads became widely available with the debut of the iTunes Store in 2003 where people can buy single tracks, not only collections. With the iPod, a whole family of products became available for different download habits (iPod Shuffle, iPod Nano, etc.).

Since 2004 Napster has been selling its 700,000 songs for USD 0.99 per track. In 2005 the US Supreme Court put an end to the illegal use of P2P technology for distributing copyright protected works. It did not stop it but made circumventing the copyrights more difficult.

Today in most countries downloads are allowed, but uploads are forbidden by law. The traditional music industry still has its problems. The Internet has shattered its business model: Today most musicians use the Internet to get famous. They do not need one of the big music companies anymore. Once a celebrity, they make their money with big concerts and fees from television. Two technologies, P2P and MP3, have killed a multi-billion industry within less than two decades.

When such revolutions bring radical changes, every company has to ask itself: How can we be on the winning side?

> There are the usual hordes of rebels and rogues, plus scores of pioneers and gold-diggers, each scrambling to carve out new territories and stake their claim in them.
> *Debora Spar* (Spar, 2001)

7.2 The Lock-in-Effect: The Winner Takes It All

An additional user of Swisscom mobile increases the value of all the phones of all the members already using the net: They can all connect to one more member! Positive feedback leads in many cases to the emergence of standards and the "winner-takes-it-all" phenomena. Unfortunately such externalities may end up in sub-optimal choices and market failure.

7.2.1 Traditional Economics

Traditional economics states: The lower the price, the higher the turnover. This is common sense but sometimes does not hold. Because of improving technology and heightened competition, prices may fall very fast within the first 10–15 years of a new product life, such as in the case of the pocket calculator. But demand picks up very slowly. In the second phase markets grow very quickly, with only a slight decrease in prices. It seems once the market has accepted a product, prices depend mostly on competition and technicalities, not on demand. There is a difference between correlation and causality.

In World War I the American bomber producers learnt that with an increase of 100 % in production, costs decreased by 20 %. Even though the theory of economies of scale was not established until the 1960s (Jansen, 2004), it had been observed before. The theory seemed to be true for many goods, such as Gillette's razors or digital watches.

The reason is mostly learning effects. The more an organization produces, the better it understands every single step of production, logistics and purchasing. This leads to an advantage of the early mover: He is farther down the experience curve than the latecomer. Together with the classical wisdom that cheaper products mean more turnover piece-wise, this leads to a classical "the winner takes it all" effect: The first to market can charge higher prices or sell a higher volume than the follower and if the first mover does not make a mistake, he will remain in the lead forever! Well, we know that this is an approximation at best: Sometimes technology changes, laws change all of a sudden. Imagine the situation of Kodak with assets in chemical factories, useless for the new digital photography. But the leading position may last for decades as in the music industry.

At the beginning of the last century Hermann Heinrich Gossen (1810–1858), a Prussian economist, formulated two laws:
1. The law of diminishing marginal utility: Marginal utilities are diminishing with increasing availability, as scarcity is a precondition of economic value. If everybody has a camera, every additional camera is less valuable.
2. The law of equal marginal utility: This law presumes that an economic agent will allocate his or her expenditures such that the ratio of the marginal utility of each good or service to its price (the optimal satisfaction with all goods available) is equal for every other good or service. But: Ferraris come in bigger chunks than yoghurts.

Both laws are not natural laws. In specific situations we may only observe patterns.

This classical situation imposes negative feedback: The utility goes down for the consumer the more he buys, and even the leader in the market will see diminishing returns the more he produces. Negative feedback leads to equilibrium for volumes between supply and demand.

Schumpeter, who sees economy as an ever-changing process driven by entrepreneurs and their innovations, has always questioned this concept. For him, economy is never in equilibrium.

Innovation is a factor that disturbs the equilibrium from time to time. Innovation economics positions knowledge, technology, entrepreneurship and innovation at the centre of economic models rather than taking these parameters as independent forces that are largely unaffected by policy. Innovation economics is based on two fundamental tenets: First, the central goal of economic policy should be to spur higher productivity through greater innovation, and second, markets relying on input resources and price signals alone will not always be as effective in spurring higher productivity, and thereby economic growth.

7.2.2 Positive Feedback Loops

As long as the market works, each technology gets its share according to its price-volume ratio.

Externalities may change the picture: Cement producers blow millions of tons of CO_2 into the air – in total some 6 % of all CO_2 produced by human activity. This has a negative effect on third parties, the public. If the public does not react, the existing cement technology will produce costs carried by the public without paying for it, a negative externality. Customers will buy cement without knowing that they pay more than what is on the bill. Clean technologies reducing overall costs could have problems being successful in the market.

When Microsoft pushed the software package Windows Explorer onto the market, it became a standard. Not because it was extremely good, but because it was a Microsoft product, the market leader. Each user not using Windows, e.g. Linux users, got an incentive to buy it as well, as the vast majority used it and got locked in the technology. Even though Microsoft did not have a better technology, Microsoft could make more money investing it in R&D resources. If Microsoft had not received positive feedback through its market position, Linux would have also had the chance to put more money into R&D. Thus positive feedback may lead to sub-optimal standards and sometimes might be seen as producing market failures.

There is one industry which is especially prone to externalities: Network IT. The reason is Metcalf's' law: The member of possible connections between two members in a network with n members is equal to $0.5*(n-1)*n$, which is approximately $0.5*n^2$ for big networks. Thus the value of a network increases with the square of its number of members. If a company is able to set up a network, it can happen that the winner takes it all.

Fig. 7.2 Blown to bits (Evans & Warster, 2000): the trade-off between richness and reach

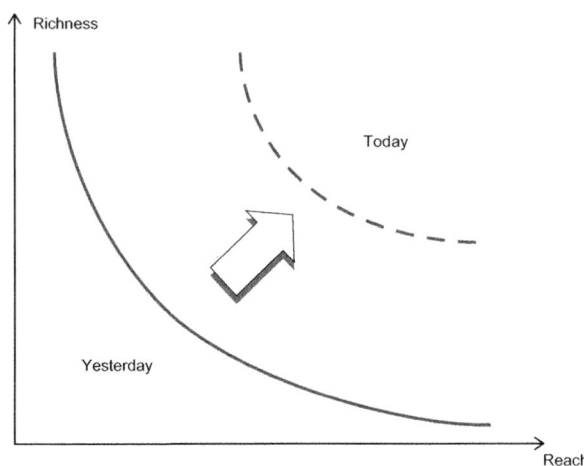

7.2.3 Effects on the Industry

Modern IT has led to an explosion of connectivity. Fifteen years ago we still had a big trade-off between reach and richness (Fig. 7.2): With some media such as radio broadcast you could reach millions of people, but the content of the message, the richness, was limited. Today we can reach far more people through the Internet and richness is much higher.

Some concrete examples of this development are:
- More freelancing and job rotations within asset management. You don't need to be an employee in a bank to get access to the information needed.
- The Encyclopaedia Britannica has moved to the Internet and is now a web portal, like Wikipedia. Reach and richness are much better on the Internet.
- Newspapers disintegrate. The most profitable parts may move to the Internet. More and more jobs are found through the Internet.
- MP3, iPod, iPhone revolutionize the market for music through smaller transaction costs.

Standards are very important in this trend. More and more standards evolve; some of them are content-driven. General Motors is the biggest buyer of health products in the United States because it has a huge number of employees and even more retired people. Standards may evolve where nobody expected it. Not the best will do, but rather "good enough" with the most power behind it. Once a standard is fixed, it can be very expensive to switch from other technologies to this standard. Thus even sub-optimal standards may survive. Another effect that points in the direction of the winner takes it all.

The potential for software applications where lock-in effects are possible is huge. Before the dot-com bubble burst in 2001, Credit Suisse First Boston (CSFB) estimated global transaction fees above USD 400 billion per year. The first targets could be growing industries that are fragmented make a profit, have big assets and a high degree of IT adoption. One example was the US food service industry. It had about 25,000 producers, some 4,000 distributors and more than 700,000 operators (restaurants, hospitals, schools and canteens). This bright outlook convinced many entrepreneurs to start their new business: The Dotcom bubble was born. And it burst. It takes far more time than consultants thought in the community prior to 2001 to set up all these software applications, train the users and allow the standards to evolve that guarantee the network capability. After the bubble burst the industry began moving in the direction CSFB announced, but with much less fanfare and some more sense for reality, and at much lower speed.

The Economist: Virtual currencies: Mining digital gold

Even if it crashes, Bitcoin may make a dent in the financial world

© The Economist Newspaper Limited, London (Apr 13th 2013)

IN 1999 an 18-year-old called Shawn Fanning changed the music industry for ever. He developed a service, Napster that allowed individuals to swap music files with one another, instead of buying pricey compact discs from record labels. Lawsuits followed and in July 2001 Napster was shut down. But the idea lives on, in the form of BitTorrent and other peer-to-peer filesharers; the Napster brand is still used by a legal music-downloading service.

The story of Napster helps to explain the excitement about Bitcoin, a digital currency that is based on similar technology. In January a unit of Bitcoin cost around $15 (Bitcoins can be broken down to eight decimal places for small transactions). By the time *The Economist* went to press on April 11th, it had settled at $179, taking the value of all Bitcoins in circulation to $2 billion. Bitcoin has become one of the world's hottest investments, a bubble inflated by social media, loose capital in search of the newest new thing and perhaps even by bank depositors unnerved by recent events in Cyprus.

Just like Napster, Bitcoin may crash but leave a lasting legacy. Indeed, the currency experienced a sharp correction on April 10th—at one point losing close to half of its value before recovering sharply (see chart). Yet the price is the least interesting thing about Bitcoin, says Tony Gallippi, founder of BitPay, a firm that processes Bitcoin payments for merchants. More important is the currency's ability to make e-commerce much easier than it is today.

Bitcoin is not the only digital currency, nor the only successful one. Gamers on Second Life, a virtual world, pay with Linden Dollars; customers of Tencent, a Chinese internet giant, deal in QQ Coins; and Facebook sells "Credits". What makes Bitcoin different is that, unlike other online (and offline) currencies, it is neither created nor administered by a single authority such as a central bank.

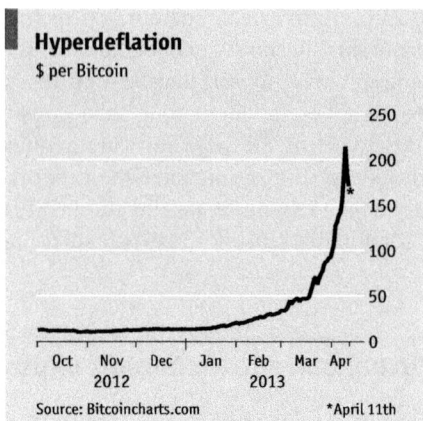

Hyperdeflation
$ per Bitcoin

Source: Bitcoincharts.com   *April 11th

Instead, "monetary policy" is determined by clever algorithms. New Bitcoins have to be "mined", meaning users can acquire them by having their computers compete to solve complex mathematical problems (the winners get the virtual cash). The coins themselves are simply strings of numbers. They are thus a completely decentralised currency: a sort of digital gold.

Bitcoin's inventor, Satoshi Nakamoto, is a mysterious hacker (or a group of hackers) who created it in 2009 and disappeared from the internet some time in 2010. The currency's early adopters have tended to be tech-loving libertarians and gold bugs, determined to break free of government control. The most infamous place where Bitcoin is used is Silk Road, a marketplace hidden in an anonymised part of the web called Tor. Users order goods—typically illegal drugs—and pay with Bitcoins.

Some legal businesses have started to accept Bitcoins. Among them are Reddit, a social-media site, and WordPress, which provides web hosting and software for bloggers. The appeal for merchants is strong. Firms such as BitPay offer spot-price conversion into dollars. Fees are typically far less than those charged by credit-card companies or banks, particularly for orders from abroad. And Bitcoin transactions cannot be reversed, so frauds cannot leave retailers out of pocket.

Yet for Bitcoins to go mainstream much has to happen, says Fred Ehrsam, the co-developer of Coinbase, a Californian Bitcoin exchange and "wallet service", where users can store their digital fortune. Getting hold of Bitcoins for the first time is difficult. Using them is fiddly. They can be stolen by hackers or just lost, like dollar bills in a washing machine. Several Bitcoin exchanges have suffered thefts and crashes over the past two years.

Ripple effects

As a result, the Bitcoin business has consolidated. The leading exchange is Mt. Gox. Based in Tokyo and run by two Frenchmen, it processes around 80% of Bitcoin-dollar trades. If such a business failed, the currency would be cut off at the knees. In fact, the price hiccup on April 10th was sparked by a software breakdown at Mt.Gox, which panicked many Bitcoin users. The currency's legal status is unclear, too. On March 18th the Financial Crimes Enforcement Network, an American government agency, proposed to regulate Bitcoin exchanges; this suggests that the agency is unlikely to shut them down.

Technical problems will also have to be overcome, says Mike Hearn, a Bitcoin expert. As more users join the network, the amount of data that has to circulate among them (to verify ownership of each Bitcoin) gets bigger, which slows the system down. Technical fixes could help but they are hard to deploy: all users must upgrade their Bitcoin wallet and mining software. Mr Hearn worries that the currency could grow too fast for its own good.

But the real threat is competition. Bitcoin-boosters like to point out that, unlike fiat money, new Bitcoins cannot be created at whim. That is true, but a new digital currency can be. Alternatives are already in development. Litecoin, a Bitcoin clone, is one. So far it is only used by a tiny hard-core of geeks, but it too has shot up in price of late. Rumour has it that Litecoin will be tradable on Mt.Gox soon.

A less nerdy alternative is Ripple. It will be much easier to use than Bitcoin, says Chris Larsen, a serial entrepreneur from Silicon Valley and co-founder of OpenCoin, the start-up behind Ripple. Transactions are approved (or not) in a few seconds, compared with the ten minutes a typical Bitcoin trade takes to be confirmed. There is no mystery about the origins of Ripple nor (yet) any association with criminal or other dubious activities.

OpenCoin is expected to start handing out Ripples to the public in May. It has created 100 billion, a number it promises never to increase. To give the new currency momentum, OpenCoin plans eventually to give away 75% of the supply. Existing Bitcoin users can already claim free Ripples and eventually anyone opening an OpenCoin account will also receive some.

The 25% retained by OpenCoin will give it a huge incentive to make sure that the Ripple is strong: the higher its value, the bigger the reward for OpenCoin's investors when the firm cashes out. On April 10th several blue-chip venture-capital firms, including the ultra-hip Andreessen Horowitz, announced that they had invested in OpenCoin.

If Ripple gains traction, even bigger financial players may enter the fray. A firm such as Visa could create its own cheap instant international-payments system, notes BitPay's Mr Gallippi. And what if a country were to issue algorithmic money?

At that point Bitcoin would probably be bust. But if that happened, its creators would have achieved something like Mr Fanning. Napster and other file-sharing services have forced the music industry to embrace online services such as iTunes or Spotify. Bitcoin's price may collapse; its users may suddenly switch to another currency. But the chances are that some form of digital money will make a lasting impression on the financial landscape.

References

Cowen, T. (2008). So we thought. But then again. *New York Times*, p. 13.
Evans, P. Y. & Warster T. (2000). *Blown to bits: How the new economics of information transforms strategy*. Boston, MA: Harvard Business Press.
Jansen, S. (2004). *Management von Unternehmenszusammenschlüssen*. Stuttgart, Germany: Klett-Cotta.
North, D. (1991). Institutions. *Journal of Economic Perspectives, 1*(5), 97–112.
Saw, C. L., & Koh, W. T. (2005). Does P2P have a future? Perspectives from Singapore. *International Journal of Law and Information Technology, 13*(3), 413–436.
Spar, D. (2001). *Pirates, prophets and pioneers*. London, England: Random Books.

Dominant Design Industry

8

> *Standards are always out of date. That is what makes them standards.*
> Alan Bennett, born 1934, English author

Abstract
Once the teething problems of a new technology are over, many companies realize its potential and experiment with different solutions until a dominant design emerges. Years or even decades later consolidation sets in and few global players survive.

In many industries we have a basic standard for whole product families on the market: We know what a car or a cell phone looks like. The similarities between devices such as refrigerators from different companies are not limited to their exterior design. The technical details which are out of sight are in many cases highly standardized. For many devices there is a technical paradigm which is a dominant design, valid for decades and one which can only be changed by a Kuhnian revolution. But vested interests and invested capital are too high to have modifications every few years. Different designs compete against each other; standard and established technologies fight for their survival and improve in their efficiency substantially only once they come under pressure from new products. This improvement extends the life of the old technology and pushes new technologies into niches at the beginning. Dominant designs lead to a crowding out of firms (Engler, 1970) (Fig. 8.1). Revolutions take longer than the newcomers expect.

A dominant design is not determined by technology alone. The more open and complex a product is and the more socially and politically questionable, the more its dominant design is influenced by society. A typical example for this interplay between technology, society and politics are nuclear power plants. Once the authorities and the public accept a design, nobody wants to change it, even if the new design would be safer.

Weight sensors are not under much pressure from the public because the sensors are hidden from the customers' eyes. They are rather influenced by their technicians. Most users do not even know what technology they are using. The

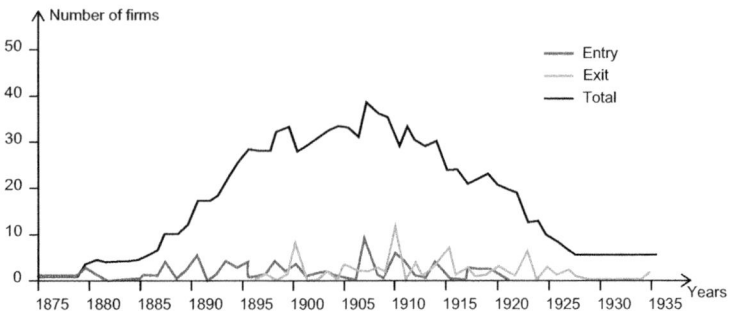

Fig. 8.1 Dominant design leads to crowding out of firms (Adapted to data of Engler, 1970)

choice of which weight sensor to use is based on technical characteristics such as precision, resolution and most importantly: cost. The pot-technology (Fig. 8.2) with its very special lever has a high precision and resolution and is therefore used in laboratories. In contrast, strings have different frequencies under different tensions with a lower precision. They are cheaper and have found their dominant design in industries and households. The cheapest technology, a strip with electrical resistance depending on how much it is stretched, is used in most scales to monitor our body weight.

The same is true for textile machinery. Width and complexity of the fabric define the dominant design of the machines. The gripper-weaving machine by Dornier has emerged as the dominant design for heavy and narrow pieces of fabric. It is well suited for high precision, but is expensive and slow. For wider fabrics the projectile weaving machine by Sulzer won recognition. Air jet weaving machines are considered as the most productive and are applicable for narrow and light fabrics.

In contrast, SMS on mobile phones became a dominant design very quickly with no market segmentation at all. No technician would have believed it, but society made the choice. SMS, short message service, is a service developed by German engineers in the 1980s. At the beginning it was restricted to 100 signs, because most postcards contained less than 100 letters! In 2012 approximately 60 billion SMS were sent in Germany alone, close to 1,000 SMS per inhabitant! Today Twitter seems to have replaced postcards. We have standards and dominant designs everywhere.

8.1 Dynamics of Dominant Designs

Abernathy and Utterback have developed a simple model to explain what happens when a new technology develops into new standard. The process consists of three phases (Fig. 8.3):

8.1 Dynamics of Dominant Designs

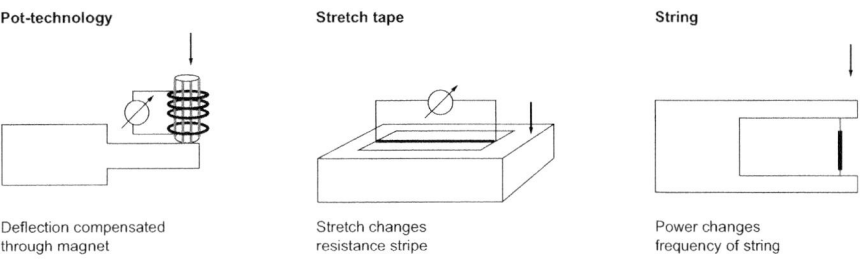

Fig. 8.2 Most important characteristic for the choice of the weight sensor: cost

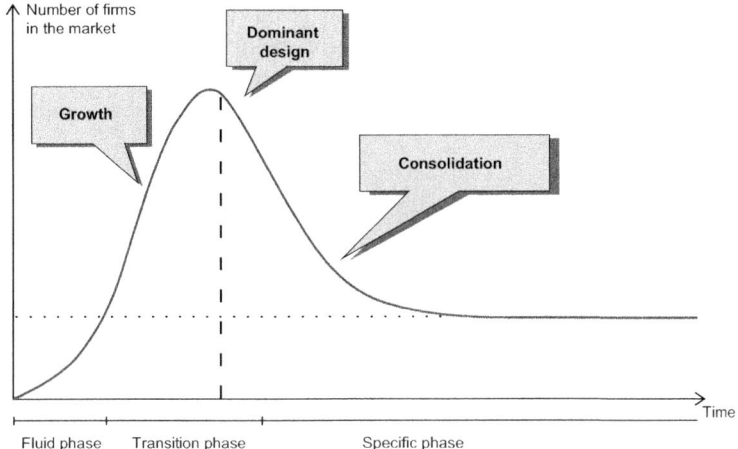

Fig. 8.3 Abernathy-Utterback-Model: the number of firms follows a "bell curve" with some firms surviving ("fat tail of the bell curve") (Abernathy & Utterback, 1978)

8.1.1 Fluid Phase

Whenever a company develops a new technology, the market spreads the rumour that this could be the next big business. Many companies try their luck. In the first phase, firms experiment and compete with different designs that improve functionality. The market is tested using many different approaches based on the users' needs and different technical concepts. Product innovation is the name of the game. The production process is flexible in order to easily accommodate major changes and lot sizes are rather small. Plants are located where they were in the past. Uncertainty looms large and competition is mostly between small and medium enterprises (SMEs). Big money is useless, risks are too high and opportunities are still too insignificant. Thus small suppliers prevail. This was the typical environment of a biotechnology start-up at the beginning of the twenty-first century or of a software start-up that tries a niche.

8.1.2 Transition Phase

Later, designs converge or disappear if the fluid phase turns out to be a bursting bubble, as in 2001 in the aftermath of the dot-com hype. The market selects some product alternatives, standards set in and volume increases fast. Due to the rising volume, process innovations are predominant. Internal technical capability is expanded and opens up many opportunities. The production process becomes more rigid. Towards the end of this phase a dominant design emerges in many cases. Prices fall and some companies already flee the heavy competition. At the peak there are at least dozens of companies on average per sector in the US alone. Around 1910, when Ford entered the car industry, there were more than 500 - car-manufacturing companies in the US. It may be that no dominant design emerges or that there are only regional dominant designs. In 2010 there is still no dominant design for houses. Architects and owners prefer prototypes. But there are dominant designs for house heating, for example gas, oil or heat pumps combined with electricity.

8.1.3 Distinct Phase

In many cases the transition phase leads to the emergence of the dominant design. This allows the "elephants", the big companies, to move in. The pressure to reduce costs and improve quality increases. It is easier for big companies to cut costs than for small companies because they are able to invest heavily in process innovation, branding and aesthetical design. The production process becomes rigid, capital-intensive and efficient; the costs of changes are high. More and more companies are squeezed out of the market. At the end there are three to five global players.

Typical examples of products in the distinct phase include cars, mobile phones and as of recently, cement producers. Sometimes the situation changes rather fast. In 2007 there were approximately 120 domestic car manufacturers in China. Most are not successful at all, but they try. This regional market is protected. We can expect that this new car economy will go through all the phases of the Abernathy-Utterback model within the next few years: Most probably only a few Chinese car manufacturers will survive the competition.

There is a lot of economic logic behind this model and it explains many of the developments we see today in the car industry, the textile machine industry and the mobile phone industry. Small companies can manage small risks and pave the way to market with innovative products. They are the "scouts" for the new market, sometimes as well as being the martyrs. Big companies adopt the successful ideas and invest their engineering, purchasing power and large amount of available capital into the new field (Fig. 8.4).

The Abernathy-Utterback-Model can be extended (Abernathy & Utterback, 1978): At the end of the early fluid phase, which is affected by uncertainty and various product variants, convergent customer needs reduce the number of alternatives of a product. This leads to companies producing the favourite products

8.2 Examples of Dominant Designs

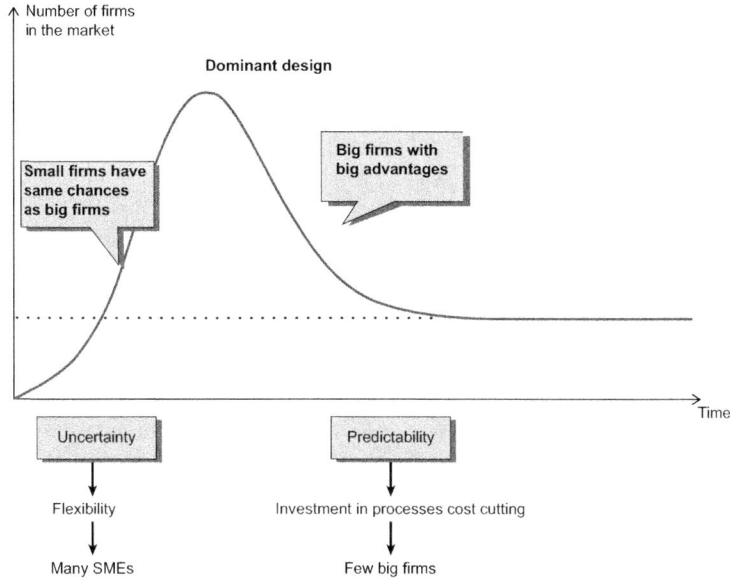

Fig. 8.4 Dominant design creates predictability and allows big investments with manageable big risks

in large quantities. Once one type of innovative product has become predominant, the transition phase begins. After the dominant design is established and big companies control of the market, a new way of competing against each other begins, involving design innovations. These sequences can be seen by means of a patent analysis in food production, for example evaporated milk. The highest number of requested patents changes from product, to processes, to design and finally to patents for specific applications (Fig. 8.5).

The theory of dominant design shows that there is a rational division of work between start-ups, SMEs and big companies: Start-ups are well suited to scout new markets. Whenever success seems possible, they merge into SMEs, make an initial public offering (IPO) or are bought by a big company. SMEs are able to fill niches and big companies know how to drive costs down and serve big markets. Small is not necessarily better than big, just different!

8.2 Examples of Dominant Designs

8.2.1 The Video Recorder

The video recorder was invented around 1950. AMPEX brought the first machine to market in 1956, the Ampex VRX-1000. It was the size of a jukebox. Then Japanese companies started to miniaturize by changing from a vertical to a diagonal trace. With the new technology the Japanese could reduce the tape width from 2 to

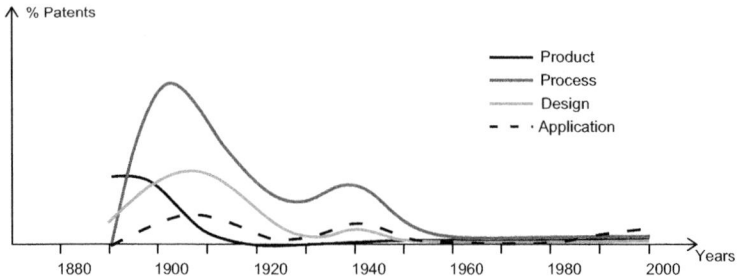

Fig. 8.5 Relative share of evaporated milk patents in relation to total patents – from product to process to design innovation (Koch & Regenscheit, 2005)

0.75 in. The hand-held video recorder became a technical possibility. After AMPEX commercialized the first video recorder in 1956, Sony, then a small company, invested 20 years of persistent R&D to make the Betamax appeal to its consumers. However, in the end its technology did not succeed as an early follower: Twenty years after AMPEX pioneered the market and at least 1 year after Sony's Betamax, the Victor Company of Japan (JVC) introduced the Video Home System (VHS), an analogue recording videotape-based cassette standard that actually succeeded because of its longer playing time, or rather the change from taping TV shows to home movie rentals. In the video recorder market, timing was not the bottleneck; the persistence needed to design a product that met customer needs was more critical.

In 1980 Japan exported more than USD 15 billion worth of consumer electronics, and 75 % of its export profits were due to video recorders. Sony thrived and today nobody remembers AMPEX for its video recorders even though AMPEX made USD 53 million with its visual storage systems in 2005. Japan set the standards and thus determined the dominant design for video recorders.

After 1980 everybody knew what a video recorder looked like, what the required quality was and how it could be embedded in an industry producing optics, tapes and cinema equipment. A dominant design was born that later integrated CCDs and memory chips. Modules changed but product architecture remained the same. Over the last few years this design has come under more and more pressure from digital cameras and cameras in mobile phones.

8.2.2 Post-It

A similar dominant design emerged over 20 years ago with Post-it from 3M. After 15 years of tinkering around and big problems in manufacturing, 3M was able to sell more than USD 2 million of post-it products in 1981. The technical challenges were enormous: How does one make a glue that sticks to two surfaces but at the end only adheres to the original side?

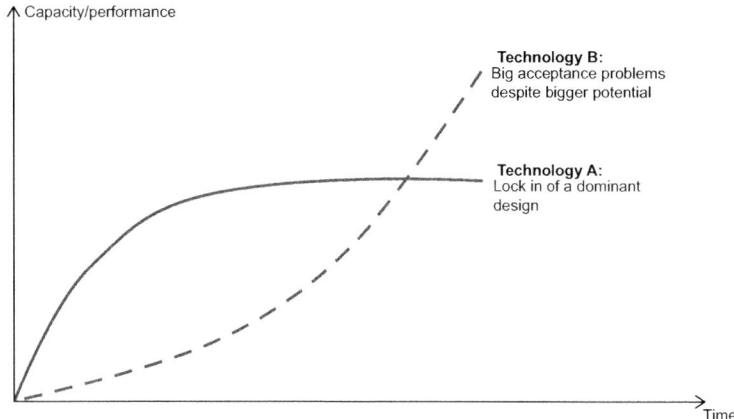

Fig. 8.6 Lock-in-effects create sub-optimal dominant designs, even in more or less free markets

3M still keeps its manufacturing processes secret. Since 1980 the number of sold post-it products has been steadily increasing. Some five million blocks with 100 sheets are sold in Germany per year, 6 sheets per inhabitant.

8.3 Sub-optimal Dominant Designs

Dominant designs are not always optimal. A typical example is the keyboard with the arrangement of letters as QWERTY or QWERZ. It is 10–15 % slower than the optimal keyboard, which would combine the fastest fingers with the most frequent letters. Because the mechanical components of the typewriter towards the end of the nineteenth century were not able to keep up with the speed of the users' fingers, Remington, one of the first producers of typewriters, had to slow down its users' typing speed and moved some frequently used letters further apart.

But once a technology is established, it sets the standard and is the dominant design. Change would need huge investments for training, production, sites and convincing end-users. The barrier to change gets higher and higher: Customers are locked in (Fig. 8.6). Retraining millions of users is not welcome with the advent of speech recognition programs which may make the keyboard redundant.

In this sense, technology has a memory. Each additional Microsoft Windows user increases the value for already existing users in terms of problem solving, additional equipment or service. The costs for switching get too high. Good enough is good enough. Thus sometimes, free markets may lead to sub-optimal technologies, sub-optimal from a technical perspective or a very long-term economic view. This is especially relevant for Internet technologies that create an "increasing marginal utility" with every new user: The more users apply a specific technology, the higher the value for every single user. Networks effects, Metcalf's law may push technical solutions into sub-optimal domains.

The Economist: The lift business: Top floor, please

Things are looking up for liftmakers
© The Economist Newspaper Limited, London (Mar 16th 2013)

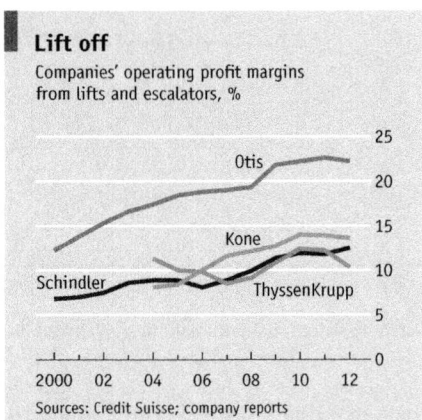

LIFTS help prove the adage that what goes up must come down. The profits of the companies that make and install them, however, seem to abide by different rules. Four firms control two-thirds of the global market: Otis, part of America's United Technologies; Kone of Finland; ThyssenKrupp, a unit of a German conglomerate; and Schindler, based in Ebikon, Switzerland. All enjoy penthouse-level margins, even as their sales glide silently upwards (see chart).

People live more vertically than ever before. An estimated 70m the world over—more than the entire population of Britain—move to cities every year. Many live in blocks of flats or work in high-rise offices. Nearly all use escalators (which the big four also make) from time to time. Few can be bothered to take the stairs.

Jürgen Tinggren, Schindler's boss, talks of a "second planet" of new consumers who are discovering the elation of elevation. Global demand for new lifts has gone from 300,000 a decade ago to nearly 700,000 this year. China, where two-thirds of new units are installed, accounts for much of the rise. Annual revenues are climbing steadily: the Freedonia Group, a consultancy, thinks they will have doubled to $90 billion in the decade to 2015.

The secret to the industry's whopping margins, however, is maintenance. People hate getting stuck in lifts. So they pay $2,000–5,000 a year to keep each one running smoothly. Since 11m machines are already in operation, many needing only a quick look-over and a dollop of grease every few months, this is a nice business. Margins are 25–35%, compared with 10% for new equipment. Revenue from maintenance is far more stable than that from installations, which are affected by the ups and downs of the economy.

A captive business
Otis and its peers make more than half their profits from services, often by securing maintenance contracts at the time of installation. This raises high barriers

to entry: a newcomer would need a dense network of technicians to get started. The incumbents do not compete ferociously on prices. In 2007 the EU antitrust authorities fined them €831m ($1.1 billion) for fixing the German and Benelux markets.

Much of the upkeep business is centred on Europe, which has around half the world's lifts (and an even higher proportion of the creaking ones that will soon need to be refurbished, another profitable niche). The market is moving east, but it will take time. "The service business comes with a big lag," says Henrik Ehrnrooth, Kone's finance chief. Some fret that Chinese lift-owners are either maintaining the machines themselves or not at all.

Schindler's Mr Tinggren says that concentrating on fixing old lifts rather than installing new ones is short-sighted. The industry must find other buttons to push; and up to a point, it has. Lifts are becoming more standardised globally, making it easier to find economies of scale. Manufacturing and R&D are increasingly outsourced to Asia, shaving costs. Sales of pricier offerings such as double-decker cabins or smarter control panels that improve the flow of people in large buildings are buoyant. Sidelines such as keycards and consulting services also pay well. New gizmos allow each lift's performance to be monitored remotely, so repairmen have to make fewer costly trips to inspect them.

Yet from the customer's perspective, lifts have evolved little since they first appeared 150 years ago. You walk in. You press a button. You avoid talking to the people standing next to you. You get off. Slim margins historically kept competition at bay. Now that the lift market is lucrative, however, it may grow more crowded.

References

Abernathy, W. J., & Utterback, J. M. (1978). Patterns of innovation in technology. *Technology Review, 80*(7), 40–47.
Engler, N. (1970). *The typewriter industry: The impact of a significant technological innovation.* Los Angeles, CA: University of California.
Koch, M., & Regenscheit, S. (2005). *Innovationsanalysen entlang des Produktlebenszyklus der Kondensmilch und deren Verpackung von 1890 bis 2005.* MAS diploma thesis, ETH Zurich.

Science-Driven Industry

9

Science means simply the aggregate of all the recipes that are always successful. The rest is literature.
Paul Valery (1871–1945), French philosopher

> **Abstract**
> Since the mid-nineteenth century technology has increasingly profited from science. The two develop in parallel: Science drives technology and technology provides more and more sophisticated research equipment. But science heavily depends on serendipity and thus many rolls of the dice are needed until a project is successful. This requires major investments and is the reason why it works better in big companies or close to universities.

9.1 Three Types of Research and Development

Technology management at Intel is certainly different from technology management at BASF, a big chemical company, and different from Mercedes where we have a whole industry developing the same products for more than 100 years and producing more than 60 million similar items per year. The question is therefore how to classify R&D situations to support management.

The Japanese Kodama has shown some interesting relations between the freezing rates of R&D projects and how tight the relation is between patents and various R&D directions (Kodama, 1995). He found three different classes of industries:

- In dominant design industries, such as the car industry, the freezing rate of R&D projects drops rather quickly and bottoms out at zero (Fig. 9.1). As the dominant design emerges it becomes an industrial standard for a specific product, providing high predictability for R&D decisions. Development projects can be frozen rather early without high risk, as the industry sticks to the dominant design and is able to invest long-term and extensively. Competition will not come up with a completely different design: Too many approval and administrative barriers would have to be overcome and even if there were a certain technical advantage, not many customers would switch product. Brands are too important. Improvements come in small steps and in general it is possible to compensate for some technical inferiority through marketing or services. An

Fig. 9.1 Freezing rate of dominant design industry drops down to zero rather quickly

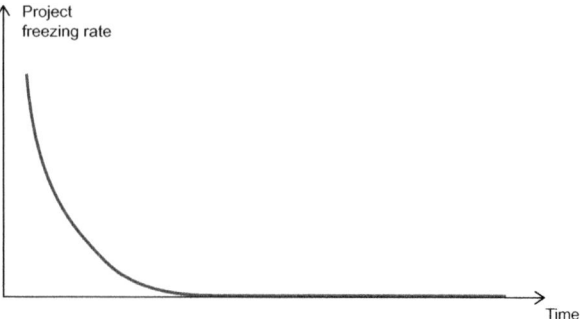

R&D project, once it has achieved certain maturity, may be carried out completely without excessively high opportunity costs. R&D is driven by efficiency.
- In high-tech industries, such as the computer industry, the freezing rate does not drop down to zero completely (Fig. 9.2); it bottoms out and remains constant, but above the zero line. Teams must be able to stop projects even at later stages when R&D expenditures are high. In high-tech you cannot compensate for inferior technology through additional efforts in marketing or production. Companies distinguish their products through better technology. For example, a 486 processor was so much better than a 386 that within a short time the market replaced most old PCs through new ones. High-tech companies are characterized by high R&D investment of more than 10 % of turnover. They employ a large number of engineers, more than 15 % of the total workforce. In high-tech industries, R&D efficiency is not enough, what is needed is effectiveness.
- In science-driven industries, such as specialty chemicals, the freezing rate is more or less constant over time (Fig. 9.3). R&D projects may be stopped in every phase, but not too often. Causes for project stops are manifold:
  1. Excessive investment needed for the final installation
  2. No approvals obtained
  3. New laws and regulations
  4. Better opportunity elsewhere

The science-driven industry is the least predictable of the three types of industries.

Kodama found that the three types of industries are closely linked to patents which he used to validate his freezing rate calculations: Science-driven industry depends heavily on patents. In Europe and in the US we have hundreds of start-ups from universities. They are transferring the newest results from science to industry. Most of these spin-offs have at the beginning nothing more than a patent and some ideas as to how to make a product and a business out of it.

Based on a study of citation data in US patents, Francis Narin developed a measure of "science linkage" which he defined as the average number of patent references to scientific papers. He felt that these references represented a company's need for scientific research (Narin & Olivastro, 1992). In a patent you

9.2 Science Versus Technology

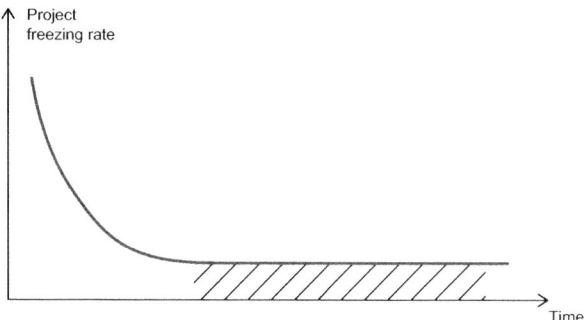

Fig. 9.2 High-tech companies must be able to stop R&D-projects even in later stages

Fig. 9.3 The freezing rate in science-driven industries is more or less constant over time

have to show you are making progress compared with existing technology. This is shown using references to old patents and references to new scientific results as well (Table 9.1).

The organic chemicals sector leads the survey in science linkage. Kodama's 177 patents in this product group cite on average 1.2 references to scientific papers. The drugs and medicine sector comes second. High-tech companies and dominant design companies have much fewer science links. The patent analysis indicates that chemicals and pharmaceuticals are science-driven industries and that high-tech industries are somewhere in between when it comes to patents. This confirms the freezing rate typology of Kodama.

9.2 Science Versus Technology

Until World War I most science was funded by private sponsors and carried out in small teams or even by individuals. In industry it was the same: Most product improvements came from small groups. One exception was Menlo Park in New Jersey, the first industrial R&D lab. It was set up 1876 by a "barely liberal, uneducated artisan, expelled from school after only 3 months for being retarded"

Table 9.1 Patent analysis gives the same typology as the freezing rate (Science link = number of citations to scientific papers per patent)

	Patents	Science link
Science-based		
Organic chemicals	177	1,2 (no Pharma!)
Inorganic chemicals	74	0,6
High-tech		
Drug and medicine	155	1,1
Communications and electronics	10	0,2
Scientific other	43	0,7
Dominant design		
Rubber	101	0,1

(Kealey, 2008). D, Development was still far away from R, Research! Also at the ETH labs at the beginning of the nineteenth century (Fig. 9.4) when systemic innovation started.

After the two World Wars it was clear the tide had finally turned. The wars showed the enormous impact of science on military equipment and thus on product development as well. Towards the end of World War II Vannevar Bush wrote his famous book: Science, the endless frontier: A report to the President, July 1945. On some 180 pages he gave support to the Baconian linear model: "Basic research leads to new knowledge. It provides scientific capital. It creates a fund from which the practical applications of knowledge must be drawn."

In 1953 the US spent some USD 40 billion on R&D with a steady increase to today's USD 360 billion, which more than doubles every 20 years!

Together with the development of ever more powerful computers the development in science has led to more and more mathematical modelling, likely culminating in the "Digital humanities" after 2013. Galileo Galilei would be happy: "Philosophy is written in that great book which ever lies before our eyes – I mean the universe. . . . This book is written in mathematical language and its characters are triangles, circles, and other geometrical figures, without whose help . . . one wanders in vain through a dark labyrinth. . ."He wrote these sentences in 1623, long before we had any computers. In his time philosophy meant science.

Even in management, mathematical models predominate despite the risks that they entail: Black Scholes' partial differential equation describes the price of a financial option over time and is the foundation of all structured financial products. These products consist of modules that have a specific risk profile. They are very similar to technical modular products. Structured products and mathematical models brought "engineering confidence" into banks and insurances: Financial engineering improves predictability and gives scientific evidence to a person's decision. It led to a boom in options trading and is widely used by options market participants because many empirical tests have shown that the Black Scholes price is "fairly close" to the observed prices. The key idea behind the derivation was to hedge options perfectly by buying and selling the underlying asset based not on human judgement but on mathematical models that should "eliminate risk". However in practice, it is widely known that blindly following models exposes the user

9.2 Science Versus Technology

Fig. 9.4 ETH laboratory beginning of the nineteenth century and beginning of systemic innovation

to unexpected risk. Among others, Black Scholes equation undoubtedly failed to help prevent the big financial crisis of 2007. Models in social sciences rely on statistics and not on natural laws: Black swans and fat tails loom large: The unexpected has low probability but is nevertheless possible!

Another example is logistics where information technology (IT) is being increasingly applied in order to manage supply chains. Big data modelling is used to predict future demand, and operation research has had a big revival. IBM is convinced that big data combined with analytical tools will become dominant in the future to manage most businesses.

Since the mid-twentieth century a more integrative view has gained increasing support (Fig. 9.5): it stresses the interdependency of science and technology and includes technology, basic and applied research, innovation, invention and diffusion. Both basic and applied research can lead to new technology. However, the opposite is also true, that new technology can lead to new discoveries, new inventions and if applied, also to new innovations.

An example that comes close to the ideal of Bacon's linear model is X-rays, discovered by Konrad Röntgen in 1895. Röntgen investigated the external effects of vacuum tube equipment and saw by chance a shadow on a small cardboard screen painted with barium platinocyanide. This was not expected nor could it be explained by any known theory. Röntgen speculated that a new kind of ray might be responsible. In the following weeks he more or less lived in his laboratory, investigating the many properties of the new rays. He temporarily termed his new discovery "X-rays", using the mathematical designation for something unknown.

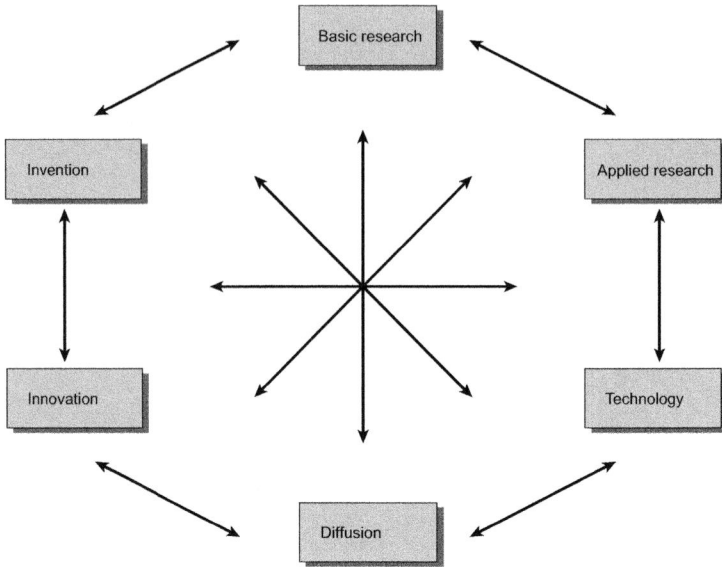

Fig. 9.5 Integrative models replace the old linear model; technology and science are highly interdependent

Two weeks after his discovery he took the very first X-ray-picture of his wife Anna Bertha's hand. When she saw her bones, she exclaimed, "I have seen my death!" Within 1 year after the detection of X-rays more than 40 scientific books were on the market, together with more than 1,000 scientific papers. X-ray technology spread with a speed that is amazing even for twenty-first century standards. Just 2 months after its discovery it was used as forensic evidence in an American court. Many scientists and engineers were severely injured by the side-effects of this new technology; nobody expected these negative aspects of a technology that had such promising advantages: For the first time in human history it was possible to look through living bodies!

Röntgen himself lived in Germany, in the middle of the German technology-cluster and science boom. He conducted research on the basis of trial and error while simultaneously pointing out the importance of mathematics.

> I was working with a Crookes tube covered by a shield of black cardboard. A piece of barium platinocyanide paper lay on the bench there. I had been passing a current through the tube, and I noticed a peculiar black line across the paper. ...
>
> The effect was one which could only be produced, in ordinary parlance, by the passage of light. No light could come from the tube because the shield which covered it was impervious to any light known, even that of the electric arc. ...
>
> I did not think I investigated. ...
>
> I assumed that the effect must have come from the tube since its character indicated that it could come from nowhere else. ... It seemed at first a new kind of invisible light. It was clearly something new, something unrecorded. ...

9.2 Science Versus Technology

Table 9.2 Science and technology are different

Science	Technology
Science analyses and looks for explanations in general terms. Orientation is given by natural laws for prediction and the periodic table for classification. An explanation is a natural law applied to specific boundary conditions	Technology is very much problem solving oriented, where problems emerge from practical applications. Technology is driven by its own problems and the need to improve performance and cost
Science lives from open questions and surprises. Every surprise is a potential candidate for a new research field and new results	Technology does not like side-effects; everything has to be controlled in order to avoid accidents and poor performance. Technology has to reach a goal and has to yield a reliable product
The goal of scientific projects usually stems from within science	The goals of technology are defined by the needs of customers. Good enough today is better than a perfect solution in the future
Scientists publish their results in global journals. Science is public and reputation is built on priority: Publish or perish!	New technologies are developed in companies and are kept secret from competitors as long as possible. Most technology is not published at all and if so, in patents only

> There is much to do, and I am busy, very busy (Describing the X-rays he had made on November 8th 1895 to a journalist.)

For Röntgen science was asking questions and gaining a better understanding through experiments and theory, always in the hope of a surprise, something new. His aim was not application, not business, but understanding and explanation. Nevertheless, his results were immediately used to develop new products by industry. To be close to universities and to science is a great advantage and has led to the science-driven industry.

Despite the increasing collaboration between universities and industry, there remain generic differences between science and technology (Table 9.2). Collaboration between a company and a research institute needs a specific management model that accepts the difference between the two.

In practice, science and technology are very much interwoven, entangled in a net where each pushes the other ahead. The first EPROM, Erasable programmable read-only-memory, was a failure in the manufacturing process of memory chips. They retained their data when their power supply was switched off. Through exposure to strong ultraviolet light, typically for 10 min or longer, data could be erased, and then rewritten with a process that needs higher than usual voltage applied. Repeated exposure to UV light would eventually wear out an EPROM, but the endurance of most EPROM chips exceeded 1,000 cycles of erasing and reprogramming. EPROM chip packages could often be identified by the prominent quartz "window" which allowed UV light to enter. After programming, the window was typically covered with a label to prevent accidental erasure. Every scientist asked for an explanation of these strange effects and was convinced that it was impossible, as it contradicted actual scientific knowledge. The first EPROM was the Intel C1701 back in 1971, which stored 256 bytes of information. Today's

EPROMs hold 8 megabyte or more. UV-EPROMs are at the end of their product life. Newer technologies are smaller, cheaper, and faster. These chips served us well into the twenty-first century, impressive for a technology that is over 30 years old. They will continue to serve in many control applications until old equipment is replaced. This is a typical example where engineers apply an effect because it works and not because science gives an explanation.

By contrast, between 1861 and 1865, based on the earlier experimental work of Faraday and other scientists, James Clerk Maxwell developed his theory of electromagnetism which predicted the existence of electromagnetic waves. In 1873 Maxwell described the theoretical basis of the propagation of electromagnetic waves in his paper to the Royal Society, "A Dynamical Theory of the Electromagnetic Field." There was no theory of electromagnetic wave propagation to guide experiments before Maxwell's treatise and its verification by Hertz and others. Luckily, Maxwell could convince Marconi to test the idea which led to the invention of the radio. Marconi was the first scientist to achieve successful radio transmission. In summer 1895 Marconi transmitted signals over 1.5 miles and in 1902 he sent the first radio message across the Atlantic even though other scientists had shown that it was impossible to send a radio signal from Europe over the Atlantic to America since the waves would propagate tangentially into space. Early experiments used the existing theories of the movement of charged particles through an electrical conductor. The ionosphere reflected the waves. Nobody understood the positive result of Marconi because the ionosphere was not yet known.

9.3 The Law of Big Numbers

Since science is very much dependent on serendipity, it must be big: Statistics only works with big numbers. Science-driven companies rely on the law of big numbers. The larger the sample size, the closer relative frequency comes to theoretical probability in relative terms, the better the predictions, at least percentage-wise. In companies with big R&D uncertainties, such as the pharmaceutical industry, managers can be specific for only a few activities in early phases of R&D. Scientists in flat structures tinker around and look for spontaneous ideas. MIT Professor Thomas Allen has investigated these R&D environments and has proven in remarkably stable results over a period of years that a decreased distance between R&D employees leads to more chance encounters with face-to-face communication and in the end to higher productivity (Allen, 1970, 2007). Scientists across all hierarchies meet by chance, remember old problems and combine them with the knowledge of their conversation partner. They need a specific type of organization (Fig. 9.6). The organizations must create an atmosphere where the unexpected and the unexplained are taken seriously. This is close to environments we find in universities: Small groups, professors with their Ph.D. students, technicians and researchers, working for many years on a specific research field.

9.4 Academic Spin-Offs

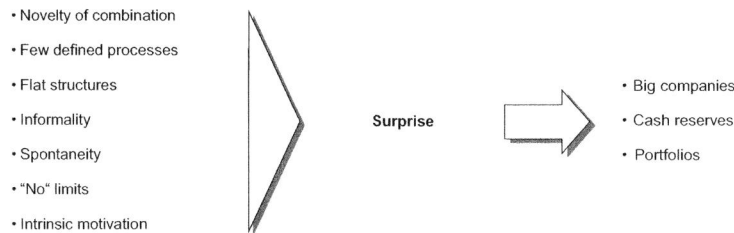

Fig. 9.6 Research: uncertainty needs specific organizations

This has consequences for the management of science-driven industries. They need to build up large reserves and a strong financial base. They need deep pockets to survive all the tortuous paths that can go on for years. R&D departments have to be closely linked to top management because research and science are the key to survival and can be measured only through peer reviews.

In the neighbourhood of these big companies and universities a new world of science-driven start-ups has evolved: Since governments are spending more and more money on their universities they would like to see a bigger "bang for their buck" as well. US universities have control over their intellectual property since the adoption of the Bayh-Dole Act in 1980. The Bayh–Dole Act or Patent and Trademark Law Amendments Act is the United States legislation for intellectual property that was paid by the federal government. It gave US universities control of their inventions. Perhaps the most important change of Bayh-Dole is that it reversed the presumption of title. Bayh-Dole permits a university, small business or non-profit institution to pursue ownership of an invention in preference to the government. Most European countries adopted a similar law in the 1980s. Governments hope to speed up the technology transfer from universities to industries.

In many cases, small start-up companies, which initially rely on the support of their universities, support the transfer: They use lab spaces, receive some initial funding and management attention. Some big companies screen these start-up companies and try to buy them as soon as breakthroughs become evident. Even this ecosystem relies on the law of big numbers: Most of the researchers worldwide conduct mainstream science for their whole working life, puzzle-solving according to Thomas Kuhn. Everybody hopes for the lucky strike! The bigger the number of start-ups, the more lucky strikes we expect.

9.4 Academic Spin-Offs

After Bayh-Dole in the US, many politicians hoped for faster technology transfer from the universities to industry and universities hoped to make more money out of these businesses. Both hopes have only been partially fulfilled.

Since academia lives from reputation, academic researchers have to publish as soon as they have a reliable result. But once published, you cannot patent anymore.

This is a first dilemma. Most technical universities have established transfer offices that help professors to overcome this dilemma. Together with financial bonuses, they hope to convince professors to patent first and wait with their publications. It seems that financial incentives are not enough: Some empirical evidence shows that professors are the bottleneck for the start of spin-offs (Schicker, 2012).

Since the lucky strike does not happen very often, financial returns to universities are low on average and highly concentrated. Just a few universities make some money. In the US one could expect USD 1–2 billion per year. For tiny Switzerland this would mean 30–50 million Swiss Francs, which is less than 1 % of the total cost of Swiss universities. Whenever big money flows back, it usually flows back to the individuals involved.

A second dilemma arises because managing R&D groups within a university is much different from management in industry or in a start-up. In industry, people have the opportunity to gain experience first as researchers, then as team leaders and later as managers. Empirical evidence shows that the single most important success factor for a start-up is the grey hair of the founder: There is not much grey hair at universities: Most researchers are young Ph.D. students or Post-Docs. The success rate of academic spin-offs is much lower than the success rate of industry spin-offs (Hess, 2012).

In the end, we should not forget that most progress in industry is due to continuous improvement in small steps. We need big, reliable and efficient players, but small, flexible and highly motivated ones as well. What the right proportion is, nobody can tell. But it is important that the administrative hurdles for start-ups are kept low for both, to start a new company as well as to stop a young company. Outside the US we have too many small start-ups that do not grow but are not closed down either.

The Economist: Global positioning: Accuracy is addictive

The invention of GPS married ideas from quantum mechanics and relativity with the need to track Russian satellites. Most remarkable of all, the concept—now the basis of a $12 billion industry—was put together over a single weekend

© The Economist Newspaper Limited, London (Mar 14th 2002)

EVER since humans took to exploring beyond their front yard, they have sought ways of determining where they are and where they are going. But until fairly recently, the best positioning tools available—landmarks, celestial guidance and dead reckoning—all had disadvantages. Some were too local or complicated; others too finicky or inaccurate. In her book "Longitude", Dava Sobel wrote that an inability to establish their whereabouts reliably caused every great captain in the age of exploration, from Vasco da Gama to Sir Francis Drake, to become lost when at sea.

During the 1950s, with their first artificial satellite in the works, the generals of the former Soviet Union were mulling over the merits of building a satellite navigation system. The KGB agents dispatched to the United States concluded that the Americans had no interest in such a system (true at the time) and had the generals' plan dismissed as a bad idea. It took the launch of *Sputnik* in 1957 to shock America into action—and to start a desperate bid to catch up. In the process, they conceived a satellite navigation system called the global positioning system (GPS), which solved the problem, once and for all, of how to locate yourself anywhere on the globe in three dimensions.

It was during the Gulf war in 1991 that the world first began to hear about GPS in a big way. Using hand-held devices rushed into production by a small firms such as Magellan, coalition troops were able to find their way around the Iraqi desert with uncanny accuracy. Since then, GPS technology has moved into public life. Today, it is being applied with ever-increasing precision to such activities as exploring the heavens or navigating the maze of streets and freeways around Los Angeles in a rented car.

It is used to track fleets of vehicles on the highways and self-steering tractors in the fields; to keep weekend yachtsmen out of trouble at sea; and to check the health of dams and bridges around the world. But the best-kept secret about GPS is its value as a clock. Time-sensitive operations associated with banking, telecommunications and even power grids have come to depend on GPS. Already the basis of a $12 billion global industry, GPS is an example of a self-perpetuating innovation: the better it gets, the more uses people find for it.

How did it come into being? Within days of *Sputnik*'s launch in October 1957, researchers around the world were able to determine the satellite's orbit by analysing variations in the frequency (ie, the Doppler shift) of the satellite's radio signal as it passed overhead. Frank McClure, then chairman of the Applied Physics Laboratory (APL) at Johns Hopkins University in Baltimore, saw the navigational possibilities of using the Doppler shift from a satellite to locate a receiver on earth.

The following year, with money from the Pentagon, scientists at APL began to develop a satellite navigation system called Transit. At the time, the nature of the ionosphere and the earth's gravity field—which can have a serious effect on radio signals—were still conundrums. But progress was swift. The United States Navy, which was looking for a way to launch ballistic missiles accurately from nuclear submarines, assumed responsibility for the system in 1958. Within six years, Transit was up and running.

Subsequently, other satellite navigation efforts were started within the American defence establishment. The Timation project at the Naval Research Laboratory was based on highly accurate clocks. Not to be left out, the United States Air Force was sponsoring a project known simply as 621B, which used a special signal to prevent jamming.

Analysing the Doppler shift of signals from Transit's half a dozen satellites allowed a submarine to determine its location to within 25 meters. But the system took up to quarter of an hour to get a fix, and it did not give continuous positioning. In 1973, the Department of Defence proposed consolidating all the satellite navigation projects into one that combined the best features of the existing systems. After some initial wrangling, the GPS concept was hammered out over a holiday weekend.

Today, GPS consists of 28 satellites orbiting 10,900 nautical miles above the earth. An upgrade programme continually adds more satellites to the constellation. A network of ground stations—in locations as far-flung as Hawaii and Ascension Island in the South Atlantic—tracks the satellites, updating their positions and time signals to ensure their accuracy.

In essence, GPS works by measuring the time it takes a radio signal from a satellite to reach a receiver on the ground. Each satellite continuously broadcasts a signal that gives its position and the time. A GPS receiver compares its own time with the satellite's time, and uses the difference between the two to calculate the distance. Taking measurements from four satellites allows the receiver to pinpoint latitude, longitude and altitude, and to correct for errors in its clock, which is not nearly so precise (or costly) as the clocks in the satellites. The problem is simply one of solid geometry: finding the point of intersection of four spheres, each centred on one of the satellites. Using four signals is better than three; and even more is better still. Modern GPS receivers pluck down as many signals as possible.

Success, of course, depends on knowing exactly where the satellites are at any given moment. That part is not all that hard, since the satellites are constantly updated with navigational data by the control stations on the ground. But accurate timing is also critical for GPS. It takes less than a tenth of a second for the signal from a satellite overhead to reach a receiver on the ground. What device could keep time that precisely? The needed innovation, oddly enough, emerged from basic research of the most arcane sort.

Not your father's watch

If you have something that vibrates with a known frequency, you have the basis of a clock. Pendulums, springs, tuning forks and piezoelectric crystals have all been pressed into service with varying degrees of accuracy. But none compares with the oscillations that occur deep within the atom. The theory of quantum mechanics says that atoms will emit electro-magnetic radiation of a very specific frequency ("resonant frequency") depending on the particular energy states that are available to that individual atom.

Remarkably, GPS calls into play Einstein's theories of special and general relativity—among the most esoteric theories of physics

In 1944, the year he won a Nobel prize for "magnetic resonance", Isador Isaac Rabi of Columbia University suggested that an atomic clock could be made using the resonant frequencies of caesium atoms. Rabi was interested in the properties of atomic nuclei, not clocks, and he never pursued the idea. But after the second world war, his former student, Norman Ramsey, improved Rabi's method, collecting his own Nobel prize for the work in 1989.

Creating atomic-time standards became a priority in both America and Britain after the second world war. Jerrold Zacharias at the Massachusetts Institute of Technology developed a practical atomic clock around the same time that Louis Essen and John Parry established the first atomic clock at the National Physical Laboratory in Teddington, west of London. Thanks to this pioneering work, "NIST-F1", America's primary time and frequency standard at the National Institute of

Standards and Technology in Boulder, Colorado, is accurate to one second in 20m years.

Each GPS satellite carries four atomic clocks synchronised to a universal standard. Each satellite transmits information at two frequencies, L1 and L2. The signals include "pseudo-random codes", which are a sequence of "up" and "down" states that look like random electrical noise. The innovation of assigning a unique pseudo-random code to each satellite allows all the satellites in the GPS constellation to use the same frequencies without jamming one another, and protects them from other interfering signals as well. As a bonus, the random code's continuous signal provides precise information about the time.

Remarkably, GPS calls into play Einstein's theories of special and general relativity—among the most esoteric theories of physics. Because of the velocity with which the satellites orbit the earth and the weaker gravitational field that they move through, the atomic clocks on board the GPS satellites advance faster than comparable clocks on the ground. If the effects of relativity were not taken into account, GPS would not work.

It was always assumed that the public would get access to GPS one day. But it was not until the early 1990s that a sufficient complement of satellites was in orbit. While the military signals were accurate to 18 metres and encrypted for exclusive use, the civilian signals were distorted deliberately to reduce their accuracy to 100 metres. President Clinton ended the restrictions on accuracy for commercial use in May 2000. Today, the public can get GPS position fixes with an accuracy of three metres to 15 metres, depending on where they are.

Actually, GPS can do a lot better than that. By cross-referencing the satellite data with a ground-based transmitter of precisely known location, the accuracy can be boosted tenfold or more. Aviation authorities around the world have been testing "differential GPS" for blind landing in fog, where approaching aircraft need to be positioned vertically and horizontally with an accuracy of ten centimetres or less. Then there is the "advanced GPS" that is used for scientific and other precision measurements such as monitoring seismic activity and ocean currents, and for remote sensing of global warming within the atmosphere. Using additional triangulation, this can pinpoint an object's position to within one centimetre.

Thanks to Sputnik

At a time when American rockets were still toppling over on the launch-pad, the idea of using satellites to navigate was as audacious as it was ingenious. To start with none, and then to send a whole constellation of satellites into space within a few decades, represents a remarkable feat of technology. Establishing the vision and the means for organising, tracking and maintaining a project of such global proportions and complexity was equally impressive.

GPS is a classic example of the process of technical evolution. Observation of *Sputnik*'s signals led to Transit, the world's first satellite navigation system. Transit sparked competing efforts to develop an even better navigational system. In turn, that lead the Pentagon to consolidate all the projects into one focused international (though American-funded) push. And out of that came GPS.

As a case history of innovation, GPS shows the importance of being able to understand the physics at the most theoretical level while also being able to envision the technology needed to use it practically, and then ruthlessly prevent roadblocks, rivalries or lack of focus to hinder progress. It is doubtful that the global positioning industry would be where it is today without the Pentagon's role as midwife. Nevertheless, the lessons learned from GPS's commercialisation apply equally to technological innovation in the biggest and smallest of companies.

References

Allen, T. J. (1970). Communication networks in R&D laboratories. *R&D Management, 1*(1), 14–21.
Allen, T. J. (2007). Architecture and communication among product development engineers. *California Management Review, 49*, 23–41.
Hess, S. (2012). *An open innovation technology transfer concept* (Dissertation). ETH Zurich, Zurich, Switzerland.
Kealey, T. (2008). *Sex, science and profits*. London, England: William Heinemann.
Kodama, F. (1995). *Emerging patterns of innovation: Sources of Japan's technological edge*. Boston, MA: Harvard Business Press.
Narin, F., & Olivastro, D. (1992). Status report: Linkage between technology and science. *Research Policy, 21*(3), 237–249.
Schicker, A. (2012). *Technology start-ups* (Dissertation). ETH Zurich, Zurich, Switzerland.

High-Tech Industry   10

> *There might be new technology, but technological progress itself was nothing new - and over the years it had not destroyed jobs, but created them.*
>
>   Margaret Thatcher (1925–2013)

Abstract
High-tech means differentiation through better technologies. It needs a special culture: Companies have to be able to stop innovation projects whatever the sunk costs are.

High-tech is a sector in which investment in knowledge is key. Typical high-tech industries include computers, electronics, aerospace and machine tools. Companies in the high-tech sector have intensive R&D and may even have to suffer large initial losses in order to gain experience (Krugman, 1994): AMD has spent hundreds of millions of dollars to catch up with Intel; and Airbus has spent billions of Euros to meet the capabilities of Boeing. Since the advantages of high-tech products can easily be seen, they are likely to get subsidized. Today's concerns about global warming have led to large subsidies for clean technologies such as photovoltaics or fuel cells.

Kodama has shown that there is a correlation between R&D and investment (Kodama, 1995). Investment is directly proportional to R&D. However, the exact type of correlation strongly depends on the industry. In high-tech industries the R&D to assets ratio may be as high as 6:1. But we have considerable investments in factories for semi-conductors, the so-called fabs, as well: A modern CPU production line may need investments above USD 10 billion.

In statistical terms, high-tech is defined in the US as an industry that spends more than 10 % of its revenues on R&D or whose workforce has a high percentage of engineers and scientists – more than 15 %. In mature technologies like Asahi glass the ratio between R&D and investment is 1:0.5. Asahi spends much less on knowledge creation but a lot on assets such as production equipment. In Europe we also have well-known examples such as Logitech that produces mice for computers, and Holcim who produces cement in the dominant design. In 2011

Logitech spent 8 % of its net sales on R&D. Holcim invests heavily in manufacturing units and much less in R&D.

10.1 From Oil to Fuel Cells?

Mobility is one of our biggest economic achievements of the last 200 years and mainly depends on one technology: The Otto engine. It was invented in Germany in 1870 and was powered by petrol. But in 1927 the world record for the fastest mile was still held by a steam-powered car. The Great Depression in the 1930s impeded the revival of steam power because the small amounts of petroleum were easier to distribute than large volumes of water. Too much had already been invested in the dominant design. Both petroleum and water distribution were up and running. However, water was just intended for horses, not for cars. At that time, most of the raw oil came from Texas. In 2010 the US was back on the top of the list of the biggest oil-producers thanks to a newly revised technology, fracking. To protect the economy from any bottlenecks in the supply of oil, the US even went to war twice in the Gulf. But the oil from Iraq is only slowly entering the world market. For this reason everybody would like to find a substitution for oil: Governments and big companies invest billions of dollars in order to reduce our dependency on oil because the biggest oil reserves are in politically unstable regions and oil is responsible for our CO_2 footprint. Today, the fuel cell is seen as one possible alternative to oil-driven energy and thus as a way of overcoming our dependency on oil. Nobody knows whether fracking will change the whole picture over the coming years: It seems that there is an abundance of gas for decades to come.

The basic technology behind the fuel cell is the electrolysis of water. It can also be seen as taking advantage of the energy which is generated from a small controlled hydrogen explosion. Simply put, hydrogen is oxidized into two hydrogen cations at the anode, one side of the cell. The hydrogen cations then diffuse through a specific barrier, an electrolyte that separates the other side of the cell, the cathode. The two electrons that separated from the hydrogen travel through an external conductor to the cathode. At the cathode, the electrons from the conductor reduce oxygen to two oxygen anions. These anions then also diffuse through the electrolyte and react with the hydrogen cations to form water. The electrons which flow from the cathode to the anode produce the electricity – or energy – that can then be used for power.

The fuel cell proved its technical feasibility many years ago in the niche of space flight and is now being developed in Japan, the US and Europe. Several designs are competing against each other; a dominant design is not yet in sight. The designs differ mostly in their running temperature which is defined by the type of electrolyte. Depending on the physical characteristics of the electrolyte, different temperatures are needed for an optimal conductivity. The different designs show improvements by factors from model to model. Nevertheless we are still a factor of 10 away from the efficiency of existing technology. A new application along the lines of a disruptive innovation is emerging at present – the replacement of batteries. Since everybody sees the huge potential the new technology has, it receives billions of dollars of funding. Small companies collaborate to develop complete systems and even big car companies set up joint ventures 10 years ago to combine their strengths (Boutellier & Böttcher, 1999).

The cost of production of the cell stacks is one bottleneck. Another major problem is the distribution, storage and transportation of highly explosive hydrogen. The fuel cell is a typical example of a high-tech development:
- It improves fast. High-tech is for organizational theory what drosophila is for biology. Every few months a new generation arises.
- It has a huge market potential. Everybody understands what the impact could be. Therefore it receives large amounts of money from government and industry alike.
- Innovation is not only needed for the product itself, but also in many processes and enabling technologies, even for new infrastructure.
- The better technology prevails. "Better" is defined in the sense of Thomas Edison: More performance per cent.
- It is close to science and is driven by engineering. The organization with the best specialists should win the race.

In Europe, the fuel cell is being developed not only within big companies but also by a small subsidiary of Sulzer, a Swiss multi-national. In Canada it is the company Ballard. They use whatever resources are available from universities and other players in the market (Boutellier & Böttcher, 1999).

The Japanese approach in the early stages of fuel cell R&D is different. MITI, the ministry of international trade and industry, develops a new vision of Japanese technology every 10 years. It usually starts with a vision of Japanese society, its needs and its demographic development. In the case of fuel cells, Japan does not directly target the most difficult market, its application in cars, but instead tries to develop stationary fuel cells for decentralized power production. Projects are derived from these specific market needs and carried out in consortia led by MITI. Companies, universities and the government join forces to develop basic technologies in pre-competitive research. Each company itself implements the product development afterwards. In this way, MITI tries to first understand the market needs and then defines targets for basic research and technology development, a demand-articulation approach (Fig. 10.1).

The Japanese are well aware that in order to get results in a reasonable time, high-tech needs whole industries to co-operate in moving ahead fast. MITI has not been very successful in the past. Nevertheless, the fuel cell shows how important it is to collaborate in high-tech industries. The leader in such consortia or joint ventures may be a government agency or a company.

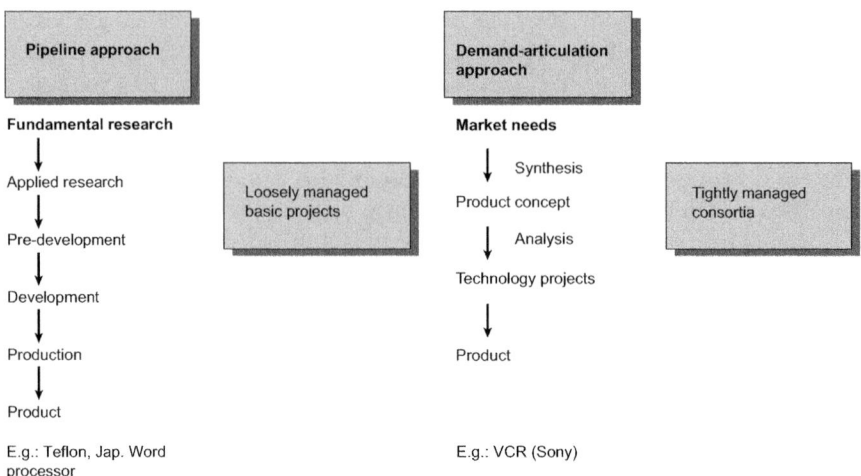

Fig. 10.1 Demand-articulation: from market needs to technology bottlenecks (Kodama, 1995)

10.2 The Microprocessor

Another well-known high-tech field is the computer industry. It has changed the world during its 60 years of existence and it has changed the structure of an entire industry from vertical competition to horizontal competition and even to co-opetition (Fig. 10.2): They collaborate in some fields and compete in others. In the beginning, large vertically integrated companies such as IBM and Siemens competed against each other with integrated products. Nowadays, there are more and more specialists who focus on a narrow module like the hard disk and compete against other hard disk producers. Along the value chain we do not see cosy partnerships but sometimes rather heavy co-opetition: Samsung is the biggest supplier of Apple and at the same time its most powerful competitor. The two fight a bitter battle over patents!

In a co-opetition, companies work closely together in order to make the common cake bigger, to grow the market segment, but may fight each other when it comes to distributing one piece.

Intel is one of the most powerful players in the value chain of computers, tablets and mobile phones. It competes with AMD, but has set the pace in CPUs for many years. It publishes its technology road map on the Internet to make sure its main suppliers are moving at the same speed. In this sense, it co-operates with AMD and is at the same time the biggest competitor of all CPU producers. Together they have lowered prices for CPUs so much that today even high performance machines use standard CPUs.

In 2006 Intel fired 10,000 employees. It seems that semi-conductors are overshooting customer needs in the basic modules like CPUs. In the 1990s most PC owners knew what type of processor they had in their computer because the

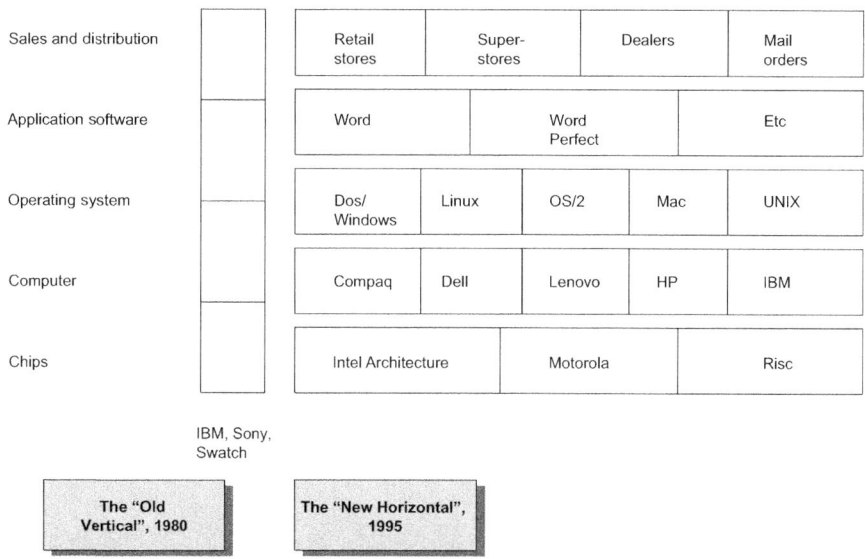

Fig. 10.2 High-tech changes industries from vertical to horizontal competition

processor was one of the bottlenecks in the computing speed and power. Today there is enough or even too much computing power, causing customers to care much less about it. Today most users are concerned about the battery lifetime: 2 or 6 h? In high-tech, technical bottlenecks may move rapidly from one module to the next.

The microprocessor industry shows that in high-tech
- co-operation may be fruitful for all market players.
- value chains disintegrate, leading to horizontal competition at the level of modules and specific steps in the value chain.
- even over-engineering may hurt. Once we have enough powerful products, customers do not pay for more power anymore.

10.3 Speed of High-Tech

High-tech has a specific culture. Whenever a competitor develops a better approach or a new technical design, the others have to follow: They have to give up old designs fast, as did the predecessors of the Pentium processor and switch to the new challenge (Fig. 10.3). You cannot sell old performance when the new generation of products is ten times better for the same price. Even though high-tech speed is generally very fast, there is the tendency in some industries for high-tech speed to slow down a bit.

Even projects in a later stage of development have to be stopped if an important competitor comes up with a better product. Top specialists leave the company,

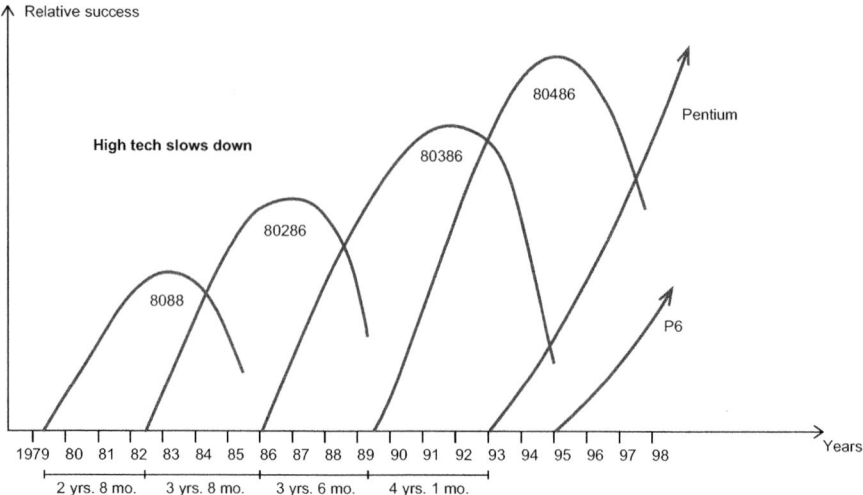

Fig. 10.3 High-tech: short lifecycle, steep ramp-up and highly rewarded leaders (but life-cycle time does not get shorter) (adapted to McGrath (1995))

company loyalty is low and technology spreads fast. Since the speed is so high, patents may sometimes be helpful, but they may be out-dated within months. No dominant design may develop. Sometimes we have dominant designs in manufacturing instead. The current standard disk size for semi-conductor production is 300 mm in 2013. Nobody knows whether this will change to 450 mm. The investments in the value chain could be in the billions.

Process technology is gaining increasing importance and needs huge investments in some cases. High-tech products have to be produced, preferably with the highest efficiency. Whenever there are positive externalities, copycats can avoid high R&D investments. The fast follower becomes a viable strategic position, providing it is possible to overcome the difficulties of the market entry and the barrier of large investments – a strategy Samsung has followed with some success in mobile phones. If companies in an industry generate knowledge that other companies can use free of charge, the industry is in effect producing some extra output which is not reflected in the incentives of the firms. Where such externalities can be shown to be important, some politicians see a case for subsidizing the industry.

10.4 Subsidies

Airbus is a knowledge-driven high-tech company. It has received billions of Euros in subsidies, 5–7 billion Euros in fact, and has reached a market share of more than 50 % after a 30 year catch-up race behind Boeing.

10.4 Subsidies

Some companies can benefit from the expensive R&D of other companies by imitating their ideas and techniques. This phenomenon can be seen in the mobile and aerospace industry, where some firms reverse-engineer their competitor's products. The current laws protect innovators only with IP rights. Incentives to innovate in fast moving markets are uncertain and some companies perceive the role of the 'copycat' as more lucrative.

The Brander-Spencer analysis is a tool to analyse such situations. It sheds some light on the battle between Airbus and Boeing. In isolation, each firm could make a profit by producing and selling aircrafts, but if both firms produce similar aircrafts and compete against each other, both will make losses, or at least much less profit. Which firm will actually make a profit? This depends on who first enters into the market. In high-tech the first mover has a big advantage. Intel is still in the lead. AMD has not been able to catch up. Let us imagine, for a moment, that Boeing gets a reasonable head start and commits itself to producing aircrafts before Airbus is able to do so. Airbus will find that it has no incentive to enter the same market. Boeing makes all the profits and is then able to survive without any help from the state. It even has the money to pay for large scale R&D. Schumpeter with his preference for monopolies would agree. However, the European government can reverse this situation by paying subsidies to Airbus so that it can enter the market.

Thus Airbus, with the help of billions of Euros in subsidies, has gained an advantage over Boeing, which means that the profits should go to Airbus. Strategic policy has changed everything. But the Brander-Spencer analysis shows that these political decisions are based on sand: The balance between the two competitors is very sensitive, even small changes in profits may change the picture. We never have enough reliable information in order to make technology forecasts or to forecast the development of markets. In the Boeing versus Airbus case, Brander-Spencer tells us how subsidies could change the rules of the game in which it could well be that the EU subsidies will never be paid back. Strategic R&D policies at the level of nations are beggar-thy-neighbour policies that increase our wealth at other countries' expense. These policies therefore risk trade wars that are disadvantageous to everyone.

Boeing has continually protested against subsidies given to Airbus, while Airbus has argued that Boeing receives illegal subsidies through military and research contracts and tax breaks. To resolve the increasing tension, subsidies have been under investigation by the World Trade Organization (WTO) since 2005. The WTO has delegated the decision to arbitration. On March 2012 the WTO stated that Boeing had received at least USD 5.3 billion in "illegal cash". But the EU paid Airbus also more than € 6 billion up to 2006. The main political argument used on both sides to subsidize high-tech firms is the creation of jobs.

Most parliaments still believe in Bacon's linear model: Subsidies support R&D, better R&D supports the high-tech industry that develops better products, which leads in turn to growth and finally to more jobs. In this belief, politicians use the following arguments:
- High-tech produces so much better products that new demand is assured
- High-tech is clean

- High-tech reduces consumption of resources
- High-tech adds high value

But high-tech might be risky as well: What about side-effects and substitutions? Subsidizing specific technologies is a risky and expensive game. What is much safer are penalties for technologies with negative side-effects: They increase costs. Engineers have been trained to reduce costs for hundreds of years. But with penalties, the choice of the technology remains with the engineers. With subsidies, politicians make the choice.

High-tech is in, especially among politicians. It has the glamour of creating jobs, of meritocracy and of being predictable. But it also creates a culture of low company loyalty and power play seen in the high-tech venture scene in the US. It was responsible for both the boom and the bust of the dot-com bubble. High-tech and low-tech are like big and small companies. We need both of them!

The Economist: The high-tech industry: Start me up

Britain has produced too few world-class technology firms. Is that about to change?

© The Economist Newspaper Limited, London (Aug 6th 2011)

MICHAEL LYNCH, chief executive of Autonomy, a Cambridge-based software firm, is showing off Aurasma, its new smartphone application. It recognises objects and then displays related video, commentary and user-created graphics around them, creating an augmented version of what the smartphone "sees". Mr Lynch hopes Aurasma will quickly capture the lion's share of a fast-growing market for web-browsing based on visual triggers rather than words. "It's a platform," he says, "so it is hard to predict its use." The more people embrace Aurasma, Mr Lynch reckons, the more businesses and advertisers will make content for it, and the more lucrative it will be for Autonomy.

This sort of virtual cycle—where users flock to a single technology "platform"—is what sets the world's best technology businesses apart. Very few of the leading outfits are British. Dozens are American. Microsoft's success rests on the ubiquity of its operating system; Intel is a giant because its personal-computer chips are the industry standard; Google is first choice for web trawlers in part because its results are derived from a greater number of searches than its rivals'; Facebook has become the global hub for social networking.

The digital economy is thriving in Britain, which spends more online per head of population than any other country, says the Boston Consulting Group (BCG). Yet it has few world-renowned firms serving the internet consumer. "In America you've had Amazon, Apple and PayPal; now there's Twitter, Facebook and LinkedIn. The whole chain is US companies," says Brent Hoberman, who co-founded Lastminute.com, a travel and leisure firm now owned by an American rival.

Only two British firms are both innovative enough to command respect in Silicon Valley, the heart of America's high-tech industry, and big enough to be in the FTSE index of the 100 largest companies. One is ARM, also based in

Cambridge, which designs the microchips that power Apple's iPhone and other portable devices. The other is Autonomy. Its Aurasma product runs on a version of IDOL, the firm's pattern-recognition software, which is used by businesses to distil order from vats of formless information—web pages, video, voice messages, e-mails and so on.

Risk and scale

That only a handful of world-class digital firms are British is more than just a matter of wounded national pride. Platform companies that can capture a digital market have benefits for the wider economy. Firms that go quickly from novice to giant create jobs and investment, of course; but they also spur growth in auxiliary trades and among local suppliers. Successful tech firms spawn start-ups, either by example or because staff leave to form other enterprises. This sort of cutting-edge innovation is closely linked to economic growth.

Britain has countless digital firms but few reach a global scale. One reason for that is the limited size of the local market. America's start-ups have 300m customers on their doorstep; Europe's market is fragmented by language and local rules, which limits prospects in ventures where Britons have a natural bent, such as retailing. Sharing a language with America can actually be a constraint. There are strong competitors to giants such as Facebook and Google in non-English language countries, says BCG's Paul Zwillenberg. Aspiring British firms have no language barrier to protect them from American competitors.

Britain has often produced great ideas but failed to turn them into big commercial ventures. That is in part down to a lack of serious commercial ambition; in the case of tech firms, it also reflects a shortage of venture capital and the support network for start-ups that comes with it. Entrepreneurs complain that local investors lack the appetite or the patience to back firms with strong growth potential, instead preferring companies with proven business models. They are envious of Silicon Valley's ecosystem of established names, serial entrepreneurs, marketing experts and specialist venture capitalists, who complement a first-rate ideas factory at nearby Stanford University.

America's appetite for riskier ventures reflects past glories as much as native boldness. Successful entrepreneurs are themselves a source of money, as are pension funds hoping for a jackpot-winning stake in the next Google. Enterprises that graduated from start-up to successful listed company in the past create the investor base for the next wave of technology stocks.

This complex is a boon to America but a drain on fledgling industries elsewhere. Promising businesses, many of them British, tend to be bought by American outfits that can offer patient capital, an industry network and access to a big market. ARM is an unusual exception: it became more valuable by not selling out, because its big customers were more comfortable locking into ARM's chip architecture if it remained a neutral party. But others need the cash and contacts more.

A recent example is Icera, a Bristol-based microchip supplier with a strong management pedigree, good technology and $250m (£152m) of venture capital behind it. Its founders set out to build a billion-dollar company that would vie with Qualcomm, the world's biggest supplier of mobile-phone chips. But in May the venture was sold for $367m to Nvidia, an America competitor.

Yet there are good reasons to hope that Britain can produce global champions. Icera may now have American owners but it is part of a mature microchip cluster in Bristol. Most of its engineers will stay there; the firm has a viable product and patents; it will spawn other start-ups. Britain's techies are adamant that they match, even surpass, their American rivals in computer science. Its engineers are good at lean hardware designs—perhaps because money to develop ideas is scarce. This brand of frugal innovation is also found in British medical-technology start-ups, says Anne Glover of Amadeus, a venture-capital firm. It suits the demands of the mobile internet where efficiency is prized.

There is growing belief, too, that the next big thing in software will come from outside America. It was once mandatory for a would-be tech mogul to seek his fortune in Silicon Valley. "To network, to find out what's going on, to meet other entrepreneurs and venture-capital people, you had to be there," says Niklas Zennström, who heads Atomico, a London-based investment firm. Today's entrepreneurs read the same blogs and tweets, and have access to a virtual community of peers, wherever they are. And America's importance as a market is dwindling: new start-ups must look to conquer emerging economies, such as Brazil, China, and India.

These currents create an opportunity for Britain, particularly its capital city. "London has an advantage, because it's the world's number-one melting pot," says Mr Zennström, who co-founded Skype, the internet-telephone company, shortly after he moved to the city in 2002. The broad outlook needed to build global companies is natural for many London entrepreneurs. It is a draw for venture capitalists, too. "We've been in London for 11 years and we've never been more bullish," says Kevin Comolli of Accel Partners, a global venture-capital firm that backed Facebook, among others. Some 30% of its European capital (including Israel) is staked on British-based firms such as Wonga, an online provider of short-term consumer loans.

Many British-based tech entrepreneurs come from abroad. Wonga's boss, Errol Damelin, is South African; Mr Zennström is Swedish. The headquarters of Badoo, a site that brings together users for casual social encounters, is in Soho, even though its founder, Andrey Andreev, is Russian, and much of the firm's early growth was in southern Europe and Latin America. London's appeal is as much social as commercial. "Entrepreneurs like the buzz and being in the thick of it," says Lloyd Price, Badoo's marketing director.

There are plenty of home-grown outfits, too, many of them based around the Old Street junction, a gateway to London's financial district. The mix of businesses in Silicon Roundabout (as it has been dubbed) fits the area's status as a creative-arts hub. Many specialise in making content that sits atop the web's existing platforms—they are as much publishing outfits as tech firms. The cluster has been championed by ministers, including David Cameron. The government hopes its cheerleading will attract investment and research to the area, helping to build a "Tech City" that stretches east to the Olympic games site.

Some tech folk are sniffy. Media start-ups with global ambitions, such as Badoo, settle in Soho, not Shoreditch, they say. Unlike the older "Silicon Fen", around

Cambridge, the London cluster has no university at its heart and (as one Bristol-based chipmaker snips) not much silicon either. But the spontaneous emergence of this group of start-ups is evidence of a vibrant entrepreneurial spirit. It is encouraging, says Jeff Skinner of London Business School, how young Londoners vie for bragging rights about their start-ups.

Could more be done to support newcomers and turn them into global leaders? Up to a point, by stimulating the flows of cash and entrepreneurs. There are too few funds to help start-ups: Cambridge Enterprises, run by the university, is a rarity. It has offices around the university, including in a new centre for entrepreneurship set up by Hermann Hauser, a Silicon Fen bigwig. Its three funds help enterprises that are too embryonic to interest venture capitalists. The outfit aims to cover costs and maintain its £8m of capital. All applicants are offered a face-to-face meeting and some initial guidance, says Bradley Hardiman, one of its investment managers. One in ten that apply gets some money; four out of five of the ventures it has supported since 1995 are still in business or have returned their initial investment.

Exile on Old Street

The government is discussing tax proposals intended to encourage more investment in early-stage ventures. But for firms to become serious players requires far bigger sums than those plans involve. Fewer smallish firms would sell out before making it big if more capital were available at the crucial growth stage, says Mike Reid of Frog Capital. A capital-gains system that encouraged successful entrepreneurs to reinvest their windfalls would also help.

As for the human capital, requiring students in relevant disciplines to write business plans would help to encourage an entrepreneurial mindset; not enough computer-science courses do so. And the supply of foreign as well as indigenous entrepreneurs could be improved. A more flexible immigration policy might produce more of them, as well as admitting more executives with the sort of digital commercial nous that is currently scarce.

But government policy can only do so much. What would help most is a few extravagant successes—to lure more venture capital to Britain's technology industry and to broaden its managerial talent. That in turn requires the ambition to become a platform company, and not just a local variant of a business model invented somewhere else. "You have to think of it as a global firm from day one," says Tudor Brown, one of the founders of ARM and now its president. If Britain is to have a global presence in technology, its entrepreneurs have to look beyond their local market.

References

Boutellier, R., & Böttcher, K. (1999). Technologien gemeinsam entwickeln. *Wissenschaftsmanagement, 3*, 207.
Kodama, F. (1995). *Emerging patterns of innovation*. Boston, MA: Harvard Business Press.
Krugman, P. (1994). *International economics* (3rd ed.). New York, NY: Harper.
McGrath, M. E. (1995). *Product strategy for high-technology companies*. New York: Richard Irwin, Inc.

Basic Innovation Principles

11

> *Genius is one percent inspiration and ninety-nine percent perspiration.*
>
> Thomas Edison (1847–1931)

Abstract

In most companies, innovation is increasingly conducted in a systematic manner. Some principles have turned out to have a significant effect on the outcome.

Innovation has become the buzzword of the managerial world at the beginning of the twenty-first century. The biggest innovation is innovation! Most people expect differentiation from competitors, application of scientific knowledge and finally growth whenever they are confronted with innovation. A closer look shows that innovation is indeed important for most companies, but has to be carried out in a systematic way in order to keep risks, cost and opportunities balanced.

11.1 Acceptance of Innovations

Innovation has to be accepted by users, customers and by employees. Thus a positive attitude and motivation are essential. This is true for both mass products and for specific single systems.

Innovation in automation, such as the development from a manually handled packaging machine to a fully automated machine (Fig. 11.1), usually has high acceptance, at least within top management.

As long as it does not destroy jobs, it is well accepted on the shop floor. In the case of biscuit packaging, the development enhanced the capabilities of the workers on the line: They became faster at producing the product, they could avoid exhausting hand movements and their working environment was improved ergonomically. In contrast to the old fully mechanical machine, the modern fully automated machine separates maintenance at the back from clean production at the front. Due to continuous improvement over more than 80 years, the acceptance of the fully automated packaging machine was never disputed neither by employees nor by customers. Experience shows that technical progress is easily accepted if it

Fig. 11.1 From a manually handled biscuit packaging machine to a fully automated modern Packaging Machine 2013 from Sigpack Systems, a Bosch Company

enhances the capabilities of those involved. However, it faces stiff resistance if it destroys existing capabilities.

Only medium acceptance was achieved for the Delta robot, which is used for picking and packaging in factories (Fig. 11.2). The Delta robot is a mechatronic device, combining mechanics and electronics, and could replace several workers at the line for packaging. Thus it met with resistance for a while.

Reymond Clavel invented the Delta robot in the early 1980s at the École Polytechnique Fédérale de Lausanne (EPFL, Switzerland).[1] The Delta robot was able to speed up the movement of individual small objects. Even though Clavel had addressed the need to improve production, he did not become successful in marketing the product and so his first start-up went bankrupt. In 1987 the company Demaurex purchased a license for the Delta robot and started producing Delta robots for the packaging industry. Reymond Clavel received the golden robot award for his doctoral thesis about his work and development of the Delta robot in 1991. In 1999, ABB Flexible Automation started selling its Delta robot, the FlexPicker. Since then, the Delta robots are also sold by Sigpack Systems, today a group of Bosch. Both Bosch and ABB have spent a lot of money industrializing the technology. The story of the Delta robot is not uncommon.

---

[1] US Patent 4976582 (A) – Device for the movement and positioning of an element in space.

Fig. 11.2 Delta robot: 150 picks per minute

If a technology changes existing processes completely, acceptance will be low at the beginning. One example is the invention of computer-aided design (CAD). It replaced the drawing board and made drawing capabilities redundant. When CAD was first introduced, engineers were much faster at their drawing boards where they had spent many years of training. With CAD they had to learn how to use a completely different program from scratch. They moved from experts to beginners! Even though CAD increased the company's overall efficiency, the individual engineer did not see any improvement.

Such different levels of acceptance always need to be kept in mind before trying to apply all the tricks of the trade to improve efficiency in innovation processes.

11.2 Eight Innovation Principles

Innovation is being carried out in an increasingly systematic way in most companies. We have seen the waterfall approach, simultaneous engineering, open innovation and now lean innovation. All these different concepts have been field-tested and are actively implemented. There is no best approach, no best practice for innovation, but some principles have turned out to have a huge effect on the outcome.

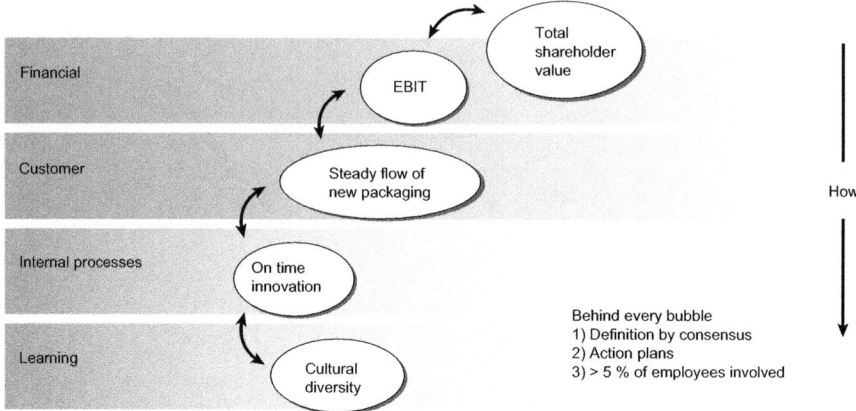

Fig. 11.3 Make strategy everyone's job: strategic road map

11.2.1 No Innovation Outside Strategy

Innovation is always a battle between creativity and discipline. Creative people tend to have their own ideas and do not easily change their direction once they have gone down one track. In order to keep them on that chosen track, it is essential that they know the strategy of the company by heart. A proven tool to achieve this is the balanced scorecard (Fig. 11.3). This may be a one–pager, a single precise report, which presents a mixture of financial and non-financial measures. One example is an internal strategic roadmap with four categories, arranged in causal order, such as (1) financial, (2) customers, (3) internal processes and (4) learning. Experience shows that all the categories and action plans are accepted when at least 5 % of all employees are involved in the strategy process. Action plans should offer a concrete way to deal with ideas in order to follow the jointly developed strategy of the company.

The strategic roadmap is easy to develop. It takes usually 2–3 workshops to set it up for the first time. The first workshop is used to define the bubbles, the second to secure the relation between the different action fields, given by the bubbles in the road map. The third workshop is usually used to fine-tune the whole action plan. The most important rule to follow in this process: Keep it simple! People have to know it by heart.

11.2.2 If It Goes Outside the Company's Strategy, Spin It Off

If the innovation process leads to products outside of the firm's strategy it is sometimes easier to spin it off than to change direction. Changing direction destroys motivation. Silicon Valley, for example, was founded by people who had walked away from their parent companies.

In 1995, SIG, a Swiss multinational company started an internal idea competition for finding new product ideas. They should be based on existing core capabilities. Four years later, in 1999, the competition was evaluated and it was found that common rail was an outside strategy and that this business had to be divested. The reasons were that too much management capacity was needed in order to develop this business. The company had a lot of experience in metal work and high pressure, but no experience with OEMs in the car business. R&D did not think ahead by building up a new sales force. Innovations need a lot of management attention: Tinkering around outside the company's strategy is in most cases more interesting than cleaning up old failures!

11.2.3 Everyone Knows the Value Proposition

Even simple products need an elaborate project organization. Technology is hidden; it is usually far more complex than the users think as well-designed products hide their technical complexity. Most people do not know that their car has software with close to 100 million program lines. To keep all of the members of the project team on the right track and to push them in the direction of a specific new product, everybody should know the value proposition of the new product by heart. The value proposition is the answer to the question: Why should the customer buy our product? As long as everybody on the innovation project wants to satisfy this question, the innovation remains customer-driven. To address customer needs for investment goods it is important that (1) customer requirements are assessed and met, (2) customer's investments are protected, (3) the payback period for the customer is as short as possible, and (4) commitments have to be made as late as possible in order to protect the company from late design changes as investment goods claim management is decisive.

The customer has to decide between different competitors and your company. So why should the customer buy your product and not the competitor's? Competitive advantages can be achieved by short delivery times, a fast service network and reliability in all aspects. To realize these features, fundamental changes may be needed in product design such as modularity, standardization and miniaturization. All have profound consequences for the organization.

It is not always easy to balance customer needs with company requirements because there are many different perspectives, not all of them known in the R&D department. We should never forget that market introduction might be far more expensive than product development. The project leader may understand the customer needs differently than the analyst or the programmer or the business consultant. The unique selling position should be unique internally as well.

Project documentation is often neglected. This can inhibit learning from feedback loops with the customer if realization takes too much time.

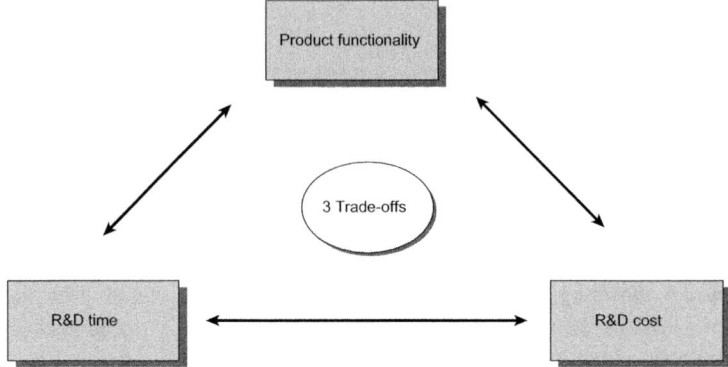

Fig. 11.4 Fix two depending on market needs, let one loose depending on customer flexibility

For mass products it is important that the value proposition addresses the needs of a whole market segment and not just one specific customer.

11.2.4 Open One Valve at the Project Start

Most innovations do not achieve the functionality which was originally planned, need twice as much time to be developed and the project costs three times more than budgeted (Fig. 11.4). This is a natural consequence of planning the unpredictable.

To avoid this surprise, top management has to define which needs have to be met in any case and where there is a degree of flexibility. Sometimes producers of machine tools develop new machines to be shown at a specific exhibition and want to achieve the same functionality as their competitors. But the R&D costs are not given. Thus management can open this valve. The project leader knows that he can get additional resources to keep up with schedule and performance targets.

Usually this procedure encounters a lot of scepticism at the top level, but using reserves is the only procedure known for coping with a surprise and surprise is everywhere in the development of new products. The valve should be opened where it hurts the least. Please do not forget that short projects tend to be cheap projects. So, opening the cost-valve is not too risky if time is kept under control.

Top management should define which of the three trade-offs or valves need to be prioritized or fixed. They can either manage this trade-off with a bottom-up or top-down approach. With the bottom-up approach, the person in charge may let the development team go forward and find out by themselves. Within the top-down approach, she can observe and react step by step or isolate the research team and give them precise orders. In the end, we have to manage the dilemma between full freedom for creativity and full discipline.

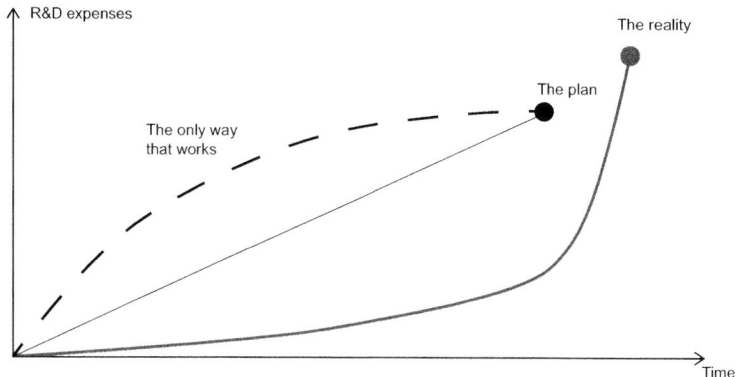

Fig. 11.5 Go for a quick start: budget overrun may be a good sign

11.2.5 Go for a Quick Start

Most project starts are a mess. People are still busy with their old projects and so the new project cannot be fully staffed from the beginning. Project members are confident that there is still enough time. Some employees do not yet believe that the new project will succeed or that it has to be finished on time. Changing such a culture is hard work. It may well be that a company has to change a few heads in R&D to achieve a new culture.

This is why, at the beginning, project costs are always below budget and increase towards the end in the typical end-race. Management has to ensure that costs are above budget from the beginning! It is better to spend too much too early than too much too late (Fig. 11.5)! This corresponds with the rule that the biggest chunk of product costs is decided in the early stages of the project.

11.2.6 Innovation Management = Project Management

In 1968, Spencer Silver, a chemist at the Minnesota Mining and Manufacturing Company (3M) in the United States wanted to develop a new super adhesive. However, he failed and only managed to develop a "low-tack", reusable, pressure-sensitive adhesive. For 5 years Silver promoted his invention within 3M, both informally and through seminars, but without much success. In 1974 Art Fry, who had attended one of Silver's seminars and was a member of a church choir, came up with the idea of using the adhesive to anchor his bookmark in his hymnbook. The post-its were permanently sticky and removable. 3M launched the product in stores in 1977 in four cities under the name "Press 'n Peel", no one saw a use for non-drying glue. A year later, in 1978, 3M issued free samples to residents and 95 % of the people who tried them said that they would buy the product. On April 6th, 1980, the product debuted in US stores as "Post-It Notes". In

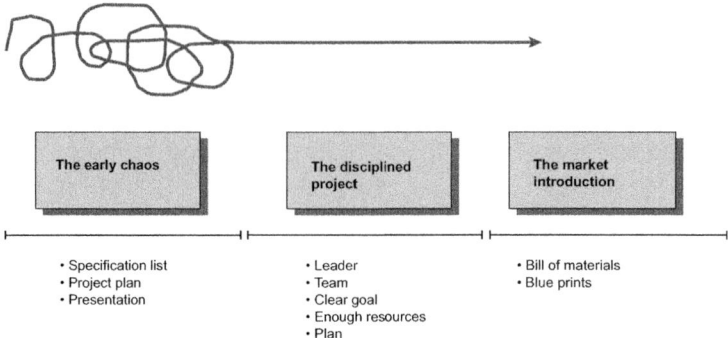

Fig. 11.6 Innovation management = project management

1981, Post-its were launched in Canada and Europe and 3M employees received the Golden Step Award for achieving more than two million sales. The yellow colour was chosen by accident; a lab next-door to the Post-it team had leftover yellow paper, which the team initially used.

Innovation projects have three typical phases (Fig. 11.6):

1. It starts with the early chaos which is vision-driven and produces two reports and a presentation: The specification list written in the language of the customer and a plan on how to conduct the project. The presentation convinces top management to start the project. It already contains a light version of a business plan.
2. The second step is the development project. A team designs the product, tries to prioritize, tries to achieve the specifications and then makes two additional plans: How to produce the product and how to market it. The bigger the project, the more important a good project leader and enough resources are. The product is documented with blue prints and a bill of materials in the case of assembled products. In pharmaceuticals the documentation may be up to a few thousand pages.
3. The third phase is the market launch. This means convincing customers and your own organization that the new product is a successful one. A generic decision has to be made: Big bang versus test market. Should we introduce the product all at once in all regions or do we test it in a specific market, improve it and only then go to the bigger markets?

The three project phases are completely different and usually need three different leaders: The first a creative geek, the second a disciplined engineer and the third a gifted and convincing salesman.

Innovation projects are similar: Only a few feasible ideas can be turned into development projects. But as soon as they are development projects, they need teams of dedicated people and disciplined project managers. After the introduction of the innovation in the market, there should be zero doubt left about the usability of the innovation. Otherwise salesmen cannot convince customers.

11.2.7 Innovation Management = Profit Management

Innovation is useful to the organization only if it makes money. That is why a business plan is a must from the beginning. In the first phase, this plan is a one-pager. It gets more and more elaborate over the course of the project. It always includes the most important trade-offs and sensitivities. At every stage of development this business plan should show a sound cash flow and a payback period that is accepted in the specific industry: Short for high technology speed, longer for low technology speed and long for lasting products. Airplane-turbines may have payback periods of 10 years and more through service and spare parts. Smartphone-projects should have payback periods measured in months.

11.2.8 Innovation Management Is People Management

The former head of R&D at Hilti, one of the top producers of hand-held construction equipment, Prof Hans-Dieter Seghezzi, once quantified R&D success in the simple formula:

$$R\&D\ success = Motivation\ *\ Qualification\ of\ R\&D\ employees.$$

We see it again and again: R&D projects succeed because of some rare individuals who fight until the end against all adversities and win. The most important individual is certainly the project leader. He knows the customer from direct contacts and knows what market segment he is targeting with his innovation. He has a project team at his hands that has all of the specific know-how needed or at least has the contacts with a network of know-how. In order for such a network to exist most teams need to be diverse and cross-functional (Allen & Henn, 2007). As several experts are now working within a cross-functional project, they need to be able to work autonomously and independently. A good team may allow for creative chaos without losing the strategic intent of the project. A project team needs analytical thinking, domain-specific knowledge and technical skills. On the other hand, they also need breadth and empathy to be able to perform in a project team (Fig. 11.7).

A project owner higher up in the hierarchy of the company should support the project leader by making fast decisions in uncertain project situations. The project owner gives his personal support needed to achieve the goals.

Innovation will always be an undertaking with many risks and many surprises but with opportunities as well. Without innovation there is no progress and more and more innovations become a routine.

We increase the R&D capacity not only to innovate, but to be able to copy as well! Most of our innovations are simply combinations of existing modules, the result of copying and exploiting market opportunities no-one else has seen before.

Fig. 11.7 Innovative thinking: generalists with a deep knowledge in at least one important field

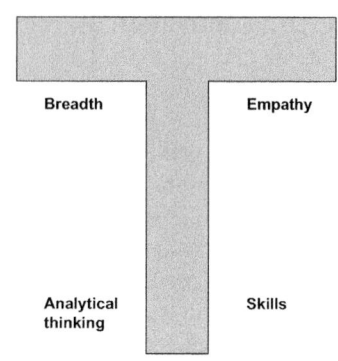

The Economist: Business innovation: Don't laugh at gilded butterflies

Rather than chasing wonder new products, big companies should focus on making lots of small improvements

© The Economist Newspaper Limited, London (Apr 22nd 2004)

THE Gillette company's website flashes out a message to the e-visitor: "Innovation is Gillette", it claims. There are few big companies that would not like to make a similar claim; for they think innovation is a bit like Botox—inject it in the right corporate places and improvements are bound to follow. But too many companies want one massive injection, one huge blockbuster, to last them for the foreseeable future. Unfortunately, successful innovation is rarely like that.

The latest manifestation of Gillette's innovative skill will appear in stores in North America next month. The global leader in men's "grooming products" is rolling out a successor to its popular three-bladed Mach3 range. It will not, as comedians had long anticipated, be a four-bladed version (Schick-Wilkinson Sword reached that landmark first, in September 2003, and Gillette has taken it to court for its pains). Rather, it will be the world's first vibrating "wet shave" blade. The battery-powered M3Power is designed to bounce around on your skin to give (yes, you guessed it) "a smoother, more comfortable shave".

For a company that claims to embody innovation, this is less than earth-shattering. On the innovation scale it falls closer to Brooks Brothers' new stain-proof tie than to the video-cassette recorder or the digital camera—especially since there is a suspicion that Gillette may be keener to create synergy between its razor and its batteries division (it owns the Duracell brand) than it is to usher in a genuinely new male-grooming experience.

But the launch is symptomatic of an important business trend: blockbuster new products are harder and harder to come by, and big companies can do much better if they focus on making lots of small things better. Adrian Slywotzky of Mercer Management Consulting says that, "in most industries, truly differentiating new-product breakthroughs are becoming increasingly rare." He claims, for example, that there has not been a single new dyestuff invented since 1956.

Even in relatively zippy businesses like pharmaceuticals, genuinely new products are fewer and further between. Spending on pharmaceutical R&D has doubled over the past decade, but the number of new drugs approved each year by America's Food and Drug Administration (the industry's key regulatory hurdle) has halved. Drug companies still live in the hope of finding a big winner that will keep their shareholders happy for a long time. But this focus means that many unglamorous, but potentially interesting, compounds may be bottled up in their laboratories.

The road to invention

Big companies have a big problem with innovation. This was most vividly described by Clayton Christensen, a Harvard Business School professor, in his book, "The Innovator's Dilemma" (Harvard Business School Press, 1997). Few conversations about innovation take place without reference to this influential work.

The Oxford English Dictionary defines innovation as "making changes to something established". Invention, by contrast, is the act of "coming upon or finding: discovery". Whereas inventors stumble across or make new things, "innovators try to change the status quo," says Bhaskar Chakravorti of the Monitor Group, another consulting firm, "which is why markets resist them." Innovations frequently disrupt the way that companies do things (and may have been doing them for years).

It is not just markets that resist innovation. Michael Hammer, co-author of another important business book ("Re-engineering the Corporation", HarperCollins) quotes the example of a PC-maker that set out to imitate Dell's famous "Build-to-Order" system of computer assembly. The company found that its attempts were frustrated not just by its head of manufacturing (who feared it would lead to most of his demesne, including his job, being outsourced), but also by the head of marketing, who did not want to upset his existing retail outlets. So the innovative proposal got nowhere. Dell continued to dominate the business.

Mr Christensen described how "disruptive innovation"—simpler, cheaper and more convenient products that seriously upset the status quo—can herald the rapid downfall of well-established and successful businesses. This, he argues, is because most organisations are designed to grow through "sustaining innovations"—the sort, like Gillette's vibrating razor, that do no more than improve on existing products for existing markets.

When they are hit by a disruptive innovation—as IBM was by the invention of the personal computer and as numerous national airlines have been by low-cost carriers—they are in danger of being blasted out of their market. This message found a ready audience, coming as it did just as giant businesses from banking to retailing, and from insurance to auction houses, were being told that some as-yet-unformed dotcom was about to knock them off their pedestal.

Innovative lessons

William Baumol, a professor at New York University, argues that big companies have been learning important lessons from the history of innovation. Consider, for example, that in general they have both cut back and re-directed their R&D spending in recent years. Gone are the droves of white-coated scientists surrounded by managers in suits anxiously awaiting the next cry of "eureka". Microsoft is a rare exception, one of the few big companies still spending big bucks on employing top scientists in the way pioneered by firms such as AT&T (with its Bell Laboratories) and Xerox (with its Palo Alto Research Centre, the legendary PARC).

This will prove to be a wise investment by Microsoft only if its scientists' output can be turned into profitable products or services. AT&T and Xerox, when in their heyday, managed to invent the transistor and the computer mouse (respectively); but they never made a penny out of them. Indeed, says Mr Baumol, the record shows that small companies have dominated the introduction of new inventions and radical innovations—independent inventors come up with most of tomorrow's clever gizmos, often creating their own commercial ventures in the process (see table).

But big companies have shifted their efforts. Mr Baumol reckons they have been forced by competition to focus on innovation as part of normal corporate activity. Rather than trying to make money from science, companies have turned R&D into an "internal, bureaucratically driven process". Innovation by big companies has become a matter of incremental improvements within the processes that constitute daily operations.

In some industries, cutbacks in R&D reflect changes in the way that new products travel down the "invention pipeline". During the late 1990s, for example, Cisco Systems kept itself at the cutting edge of its fast-moving high-tech business (making internet routers) by buying a long string of creative start-ups financed originally by venture capital. The company's R&D was, as it were, outsourced to California's venture capitalists, who brought together the marketing savvy of a big corporation and the innovative flair of a small one—functions that were famously divorced at AT&T and Xerox.

These days there is less money going into venture capital, and a new method of outsourcing R&D is on the increase. More and more of it is being shifted to cheaper locations "offshore"—in India and Russia, for example. One Indian firm, Wipro, employs 6,500 people in and around Bangalore doing R&D for others—including nine out of ten of the world's top telecom-equipment manufacturers.

Pharmaceutical giants continue to get their hands on new science by buying small innovative firms, particularly in biotech. Toby Stuart, a professor at the Columbia Business School in New York, thinks that this shows another change in the supply chain of invention. He says that many of the biotech firms are merely intermediating between the universities and "Big Pharma", the distributors and marketers of the fruits of academia's invention. Universities used to license their inventions to these firms direct, but small biotech companies make the process more efficient. They are well networked with the universities, in whose "business parks" they frequently locate their offices. They may not, of themselves, be very innovative.

Companies need to resist the feeling that it is not worth getting out of bed for anything other than a potential blockbuster. Product cycles are getting shorter and shorter across the board because innovations are more rapidly copied by competitors, pushing down margins and transforming today's consumer sensation into tomorrow's commonplace commodity. Firms have to innovate continuously and incrementally these days to lift products out of the slough of commoditisation. After it used innovation to create a commoditised market for fast food, McDonald's struggled before recently managing to reinvigorate its flow of innovations.

Finding a niche

Another factor to take into account is the fragmentation of markets. Once-uniform mass markets are breaking up into countless niches in which everything has to be customised for a small group of consumers. Looking for blockbusters in such a world is a daunting task. Vijay Vishwanath, a marketing specialist with Bain, a consulting firm, says that Gillette's bouncy blade may yet end up as no more than a niche product—fine if it is profitable.

Mr Chakravorti believes that the problem lies with the marketing of new innovations. It has not, he says, caught up with the way that consumers behave today. "Executives need to rethink the way they bring innovations to market." Too many are still stuck with the strategies used to sell Kodak's first cameras almost 120 years ago, when the product was so revolutionary that the company could forget about competition for at least a decade. Today, no innovation is an island. Each needs to take account of the network of products into which it is launched.

Companies that fail to come up with big new headline-hitting blockbusters should not despair. There are plenty of other, albeit less glamorous, areas where innovation can take place. Management thinkers have identified at least three. Erik Brynjolfsson of the MIT Sloan School of Management, says that the roots of America's productivity surge lie in a "genuine revolution in how American companies are using information technology". Good companies are using IT "to reinvent their business processes from top to bottom".

Reinventing, or simply trying to improve, business processes can offer surprising benefits to firms that do it well. The software that runs many business processes has become an important competitive weapon. Some business processes have even been awarded patents. These are controversial and, because they may stifle rather than encourage the spread of new ideas, are probably not in the wider public interest. Yet Amazon obviously views its patent for one-click internet purchasing as valuable, and there are plenty of other examples, particularly in the financial-services industry.

Nevertheless, there is no doubt that, patented or not, what Mr Hammer calls "operational innovation" can add to shareholder value. In an article in the April issue of the *Harvard Business Review*, he asks why so few companies have followed the examples of Dell, Toyota and Wal-Mart, three of the greatest creators of value in recent times. None of them has come up with a string of revolutionary new products. Where they have been creative is in their business processes.

While superficially mundane, Wal-Mart's pioneering system of "cross-docking"—shifting goods off trucks from suppliers and straight on to trucks heading for the company's stores, without them ever hitting the ground at a distribution centre—has been fundamental to the company's ability to offer lower prices, the platform for its outstanding success. Is it not over the top, though, to glorify such a common-sense change with the title "innovation"? For sure, it does not call for a higher degree in one of the obscurer corners of science. But Wal-Mart did something no competitor had ever dreamed was feasible and that was highly innovative.

Mr Hammer, who was once a professor of computer science at MIT, believes that the best qualification for innovation is a basic training in engineering. Crucially, he says, engineers are taught that design matters; that most things are part of a system in which everything interacts; that their job is to worry about trade-offs; and that they must continually be measuring the robustness of the systems they set up. Such a frame of mind, he believes, fosters innovation. It may be no coincidence that many of the greatest corporate leaders in America, Europe and Japan, past and present, trained first as engineers.

Companies are being encouraged to embrace other forms of innovation too. In a recent issue of the *MIT Sloan Management Review*, Christopher Trimble and Vijay Govindarajan, two academics from Dartmouth College's Tuck School of Business, recommend that they try a little "strategic innovation". The authors point to examples such as Southwest Airlines, a low-cost American regional carrier, and Tetra Pak, a Swedish company whose packaging products are handled at least once a day by most citizens of the western world. Such companies succeed, they say, "through innovative strategies alone, without much innovation in either the underlying technologies or the products and services sold to customers."

Tetra Pak's strategic innovation involved moving from the production of packages for its customers to the design of packaging solutions for them. Instead of delivering ready-made containers, the company increasingly provides the machinery for its customers to make their own packages: the fishing rod, not the fish.

But customers can then use only Tetra Pak's own aseptic materials to make their containers. This strips out all sorts of transport and inventory costs from the production process, for both Tetra Pak and its customer. It also makes it very difficult for the customer to switch suppliers.

Southwest's innovative strategies include its bold decision to increase capacity in the immediate aftermath of September 11th 2001, and its carefully timed rolling out this May of competitively priced routes focused on Philadelphia, an important hub for the ailing US Airways, an airline lumbered with an expensive legacy (such as highly paid crews). The low-cost carrier "is coming to kill us," said US Airways chief executive David Siegel shortly before his recent resignation. And he was not exaggerating.

In his recent book, "How to Grow When Markets Don't" (Warner Books, 2003), Mr Slywotzky and his co-author Richard Wise recommended another form of innovation. "A handful of far-sighted companies", they claim, have shifted their focus from product innovation to what they call "demand innovation". They cite examples such as Air Liquide and Johnson Controls, which have earned profits not by meeting existing demand in a new way but "by discovering new forms of demand" and adapting to meet them.

The French company Air Liquide, for example, was a market leader in the supply of industrial gases. But by the early 1990s gas had become a commodity, with only price differentiating one supplier from another. As its operating income plunged, Air Liquide tried to behave like a far-sighted company: it almost doubled its R&D expenditure. However, it reaped few fruits. An ozone-based alternative to the company's environmentally unfriendly bleach for paper and pulp, for example, required customers to undertake prohibitively expensive redesigns of their mills.

The company's saviour came serendipitously in the form of a new system for manufacturing gases at small plants erected on its customers' sites. This brought it into closer contact with its customers, and led it to realise that it could sell them skills it had gained over years—in handling hazardous materials and maximising energy efficiency, for example.

After exclusively selling gas for decades, Air Liquide became a provider of chemical- and gas-management services as well. In 1991, services accounted for 7% of its revenues; today they are close to 30%. And because service margins are higher, they account for an even bigger share of profits. An ozone-based bleach could never have done half so well.

The dilemma solved?

In his latest book, "The Innovator's Solution", published late last year, Mr Christensen argued that established companies should try to become disruptive innovators themselves. He cites, for example, Charles Schwab, which turned itself from a traditional stockbroker into a leading online broker, and Intel, which reclaimed the low end of the semiconductor market with the launch of its Celeron chip.

There are, says Mr Christensen, things that managers can do to make such innovations more likely to happen within their organisations. For example, projects with potential should be rapidly hived off into independent business units, away from the smothering influence of the status quo. The ultimate outcome of any one disruptive innovation may still be unpredictable; the process from which it emerges is not.

In the end, though, "no single innovation conveys lasting advantage," says Mr Hammer. In the toys and games business today, up to 40% of all products on the market are less than one year old. Other sectors are only a little less pressured. Innovation and, yes, invention too, have to take place continually and systematically.

Reference

Allen, T., & Henn, G. (2007). *The organization and architecture of innovation.* Burlington, Oxford: Butterworth-Heinemann and Architectural Press.

Radical Versus Routine Innovation

> *A railroad through new country, i.e., country not yet served by railroads, as soon as it gets into working, order upsets all conditions, all production functions within its radius of influence; and hardly any 'ways of doing things' which have been optimal remain so afterward.*
>
> Joseph Schumpeter (1883–1950)

Abstract

Innovation is becoming a must for most companies. And it is much easier today to innovate than it was 50 years ago, at least in small steps. Radical innovations are difficult; many are hidden. You realize their revolutionary character only in hindsight. They are not planned.

Joseph Schumpeter, father of the "disruptive storm of innovations", tried to understand the business cycles of capitalism. For him, "changes in the methods of supplying products" were the one big driver of growth and decline in the evolution of economic systems. He was convinced that every company that stopped innovating would die a "natural" death (Schumpeter & Fels, 1939).

Schumpeter's "changing the methods of supply" means a broad range of events: It includes the introduction of new commodities, which may even serve as a new standard. The definition of innovation is very broad and covers everything from a technological change in the production of the commodities already in use, creating new markets, gaining new supply sources, re-engineering work processes, improving materials handling or building new business organizations such as department stores. In short, innovation can be defined as "doing things differently" in economic life. It is not synonymous with "invention".

First, innovation seen as a natural follower of invention suggests a limitation which is most unfortunate. It is entirely immaterial whether an innovation implies a preceding scientific novelty or not. Although most innovations can be traced to some discovery in the realm of either theoretical or practical knowledge, there are many which are simply a recombination of known ideas. Innovation is possible without anything we would define as invention and invention does not necessarily lead to innovation. Invention has no intrinsic economic effect. The Stirling engine was invented more than 100 years ago. It seems that nobody has been able to create business with this special engine.

Second, even where innovation builds on a particular invention, inventing and using the invention to innovate are two entirely different things. The same person often performs both, but this is merely a coincidence. Personal attitudes – primarily intellectual in the case of the inventor, primarily volitional in the case of the businessman who turns the invention into an innovation – and the methods used belong to different spheres. The social processes which produce inventions and the social processes which produce innovations do not have any fixed relation to each other and the relation they do display is much more complex than it appears at first sight (Schumpeter & Fels, 1939).

Schumpeter has five types of innovation in mind. Today we would call them innovations of

1. Products (new commodities)
2. Processes (changes in production)
3. New markets
4. New suppliers
5. New business organizations

These five fields of activity are targets of improvement in most companies: The VW Golf of 2005 differed slightly from the 2004 model, it was produced with somewhat newer processes in an adapted organization, sold to new markets and had some new suppliers as well. We are so accustomed to these continuous improvements that we do not even call them innovations any more. Only sales people pretend that the 2005 VW Golf model is much better than its predecessor. Innovation has become a routine! Maybe VW will introduce a fully electric version of its Golf in 2020. This would be perceived as a radical innovation, because

- The product's attributes change drastically: Top acceleration, no changes of speed, no petrol ...
- It needs batteries, which would change the assembly lines, needs new suppliers and needs a special infrastructure to supply recharged batteries.
- It will appeal to different markets, different customers.
- The organization will certainly be changed as well: Perhaps batteries will be leased and not sold – a new business model is created.

12.1 Radical Innovation

Radical innovations aim to create fundamentally new businesses, products, processes and new combinations thereof. They lead to dramatic changes in the prevailing technology and are driven by new paradigms. For the company, radical innovations require a large-scale internal overhaul because they need new competencies and resources and are in many cases not only competence-destroying, but also may be even company-destroying, as in the case of Kodak. Externally, they may involve large technical advancements and make existing products non-competitive or even obsolete. Therefore according to Schumpeter, they have a revolutionary and disruptive character. MP3 together with peer-to-peer technology changed the whole music industry! Radical innovations lead to new benefits for the user unknown before. New market segments and customers need to be convinced of the new products or processes. This leads to high uncertainty and risks, as neither the application of the new technology nor its demand and economic success can be forecast. Market research does not help: Customers cannot provide information about products that do not exist. Besides fax, mobile phones, internet and e-mail, examples of radical innovations are leasing instead of investing, the container as a means of transport or the credit card as a payment innovation.

12.2 Routine Innovation

Routine innovation takes place within an already existing range of products, services, businesses or processes for already existing customers in existing market segments. The R&D department uses historical experience and applies known problem-solving methods. Typically, quality management or continuous improvement leads to routine innovation, such as performance improvement or product differentiation and adaptation to specific customer requirements. Routine innovations are mainly based on an existing innovation process that can be continuously improved in order to get new products faster with a higher quality and at lower R&D cost.

Thus routine innovations are best suited for competing in existing markets and creating short-term solutions. For a company itself, routine innovation builds upon existing knowledge and resources. It is competence-enhancing. Externally, it involves modest technological enhancements to keep existing products competitive. Surprising the customer is no goal. In general, routine innovations are easier to realize than radical innovations. They are less risky and they can be planned. A marketing department neither has to promote the product aggressively nor does it have to vie for the attention of potential customers. An example is the change from a 4 Mbit chip to a 1 Gbit USB chip.

Radical and routine innovations build on different concepts: Whereas radical innovation explores new products and processes, routine innovation exploits existing knowledge (Table 12.1) (Leifer, McDermott, & O'Connor, 2000).

As project time lines for routine innovations are short, companies can incorporate market and customer feedback regularly and significantly reduce market and financial risks. In practice most innovations are routine due to lower levels of uncertainty and easier processing. By accumulating many routine innovations they can have the same final impact as one radical change. But radical innovations also have advantages, such as fast breakthroughs in technologies, products and business in the long-term: The first VW Golf MK 1 reached the market in 1974 and the sixth model, MK 6, was shown at the Paris Car Exhibition in 2008. If the MK 6 had been presented in 1975, the whole car-world would have seen this as a very, very radical innovation! Thus the one big difference between radical and routine innovation is the speed of realization and surprise.

The great impact of radical innovations on revenue and profits was shown in a very positive light in a study by Kim and Mauborgne who are the authors of the

Table 12.1 Comparison of radical and routine innovation

Key characteristics	Radical	Routine
Main change	New businesses, products, processes that change the market	Cost or feature improvements in existing products, services and processes for current markets
Degree of novelty	High (Exploration)	Low (Exploitation)
Impact on business	Goal: creation of a rapid growth in a new market; long-term effect	Keeps companies competitive; short-term effect
Uncertainties, risks	High	Low
Project timeline	Long-term, 10 years or more	Short-term, 6 months to 2 years, depending on market segment
Trajectory	Sporadic and discontinuous	Linear and continuous
Business case	Discovery-based learning	Planned problem-solving
Process	Informal and flexible at early stages with uncertainties, after reducing uncertainties more formal	Formal, phase-gate model (concept, development, testing, marketing, manufacturing, launch)
Organization	Project starts in R&D and is transferred within the organization.	Cross-functional project team within a business unit
Resources and competencies	Competencies have to be acquired, from internal and external sources	Standard resource allocation, core competencies (Budget)

best-selling book "Blue Ocean Strategy" (Kim & Mauborgne, 2005). Please bear in mind that they compare only successful innovations! Apple tried some radical ones that were a disaster!

Kim and Mauborgne studied the business launches of 108 companies and showed that only 14 % of the innovations were radical in nature. However, these radical innovations were responsible for 38 % of the revenue and even 61 % of the profit for companies. Consequently, radical innovations, if successful, are a goal to sustain the long-term success of companies.

Thomas Kuhn and his theory of scientific revolutions and paradigms confirm that revolutions rarely occur (Kuhn, 1965). His theory can be transferred to the innovation process in companies (Fig. 12.1):

In routine phases of the innovation process (Kuhn: phases of 'normal science') everyone tries to solve the puzzles according to the existing paradigm, a well-proven world view, the hidden R&D-strategy. This existing paradigm implicitly provides rules that are unconsciously used by the engineers in the innovation process. The engineers use already existing parts of the puzzle and combine them differently – a routine innovation. The longer this paradigm exists, the more efficiently and standardized problem-solving processes can be executed by engineers. Engineers do not question basics, they take them for granted. No generic discussions arise, which makes the process very efficient. It does not make sense to question the gearbox as a whole whenever a new model for the smart is developed. It may be useful to discuss whether it should have five gears or six gears. A radical

12.3 Examples of Radical and Routine Innovations

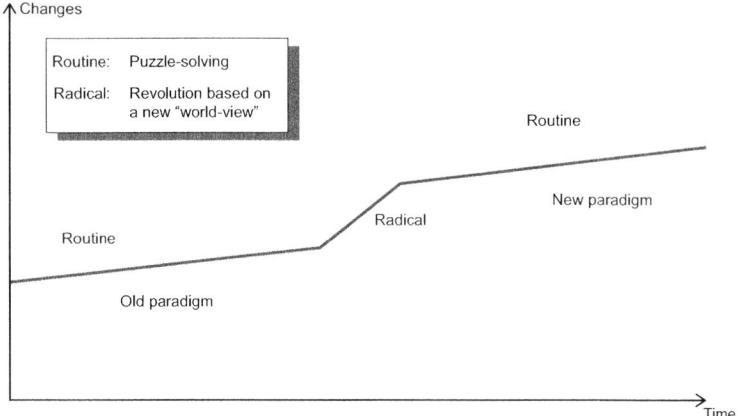

Fig. 12.1 Routine innovation represents an established paradigm

change may occur only if problems arise too often or surprising discoveries are made (Kuhn: phase of 'extraordinary science'). Toyota is successful with its "clumsy" hybrid cars and users want to have clean technology even if it is much more expensive. However, this change needs to be accepted by the engineering community. If a lot of engineers are convinced of a radical new solution, a radical innovation is born: In 2012 Audi announced its A8 hybrid at a price of over 100,000 CHF. With time, radical innovations evolve into a new paradigm: In 2012 Toyota already had its third generation of hybrids on the market. Now, engineers use the new paradigm as a basis for routine innovations. Consequently, radical is different from routine innovation but not necessarily better!

12.3 Examples of Radical and Routine Innovations

There have been many radical innovations in history. Depending on their process and purpose, their nature is different. August von Hayek, Nobel Prize Laureate, is convinced that all the very great innovations of humanity just happened and were not planned at all. Money, the alphabet, banks, hospitals or our division of power in democracies were not the result of a conscious process. But there are some remarkable innovations that were perceived as radical, even in foresight.

12.3.1 Float Glass: Dedicated Engineering

Float glass is a sheet of glass, which is produced with the float glass process or Pilkington process, named after the British glass manufacturer Pilkington who pioneered the technique. The process has been in use since the 1960s and is today used to produce about 95 % of all flat glass such as windowpanes, car glass and mirrors.

Until the nineteenth century sheets of glass were made by blowing large cylinders, cutting them open and then flattening them or producing bottles and then use the bottom only. These small sheets of glass were used for making windows for buildings and are very much loved by architects because they make a façade of a building much livelier than today's perfectly flat windows.

Even though Henry Bessemer, an English engineer, was able to patent the first automated flat glass manufacturing process in 1848, his process was expensive as the glass needed to be polished intensively. Attempts were made to form flat glass on a molten tin bath in the US, but the process did not work. More than 100 years later, between 1953 and 1957, Sir Alastair Pilkington almost went bankrupt while developing the first successful and commercially viable process for continuously forming sheets of flat glass. He used a molten tin bath onto which the molten glass could flow under the influence of gravity. The success of this process lay in carefully controlling the volume of glass fed onto the bath, where it flattened under its own weight. That is why float glass is not fully flat and why its curvature arc equals the earth's curvature.

Once the mixture is molten, the temperature of the mixture is stabilised at a temperature of approximately 1,200 °C in order to ensure a homogeneous specific gravity. The molten glass is fed into a bath of molten tin. The glass flows onto the tin surface forming a floating sheet with perfectly smooth surfaces and an even thickness. As the glass flows along the tin bath the temperature is gradually reduced from 1,100 °C to 600 °C until the sheet can be lifted from the tin onto rollers. Variation in the flow speed and roller speed enables glass sheets of varying thickness to be formed. After being cooled down, the glass is cut by machine.

Pilkington introduced the process in 1959. Full-scale profitable sales of float glass were first achieved in 1960. From then, the company Pilkington Brothers in St. Helens, Great Britain, sold licences to other float glass producers.

The life cycle of a float glass facility is approximately 15 years. An average facility is active 24 h a day and produces 3,000 m^2 of glass with a thickness of 4 mm per hour, or 33 t glass/h. About 250 float glass facilities are active worldwide.

With the automated glass manufacturing process, the prices of float glass have fallen significantly. Architects can use float glass for various purposes in buildings such as energy saving constructions with high transparency. For architects, this has changed the layout of buildings completely.

Float glass is a radical innovation: Today it is possible to produce sheets of glass, which are much bigger and much cheaper than the old glass roundels. The old production process has disappeared. However, if we take a closer look at this innovation we can agree with Schumpeter: There is not a lot of invention in this radical innovation. It is a new combination of existing technology. But to get this combination to work, Pilkington needed thousands of engineering hours and an enormous amount of time to optimize the whole process. Radical innovation is very often a novel combination of existing, slightly adapted knowledge.

12.3.2 Containerization: Global Standards

During World War II unloading and reloading ships took a long time and was expensive due to theft and damage in transit. So the US army started to experiment with containers on ships. In 1952 the Container Express (CONEX) was born. CONEX was used for shipping supplies and spare parts from Columbia to San Francisco, to Yokohama (Japan) and to Korea during the Korean War. With CONEX the transportation time was reduced from 55 days to 27 days (Miller, 2006), which was decisive for military success over long distances.

The first purpose-built container ship was planned and built by Clifford J. Rogers in 1955 in Montreal. This ship belonged to the first intermodal transport unit where containers were changed at least once from ship to truck or railroad cars.

Malcolm McLean, an American trucking entrepreneur, is known as the "father of containerization". The novelty of his innovation was that the large containers were never opened in transit and were transferable on an intermodal basis. The challenge was to design a shipping container that could efficiently be loaded onto ships and held securely on long sea voyages.

During containerization's first 20 years, many container sizes and corner fittings were used; there were dozens of incompatible container systems in the U.S. alone. In the years from 1968 to 1970 global ISO-Standards were introduced (Rushton, Oxley, & Croucher, 2004). The best-known and most important type of intermodal container for merchant shipping has the following measures: $12.19 \times 2.44 \times 2.60$ m. According to Kuhn, a radical innovation established itself and became standardized.

The paradigm of containerization has revolutionized cargo shipping. As of 2009, containers stacked on transport ships move approximately 90 % of non-bulk cargo worldwide. In 2005 approximately 18 million containers made over 200 million trips per year (Levinson, 2010).

As of today, two thirds of cross-border trades are executed using container ships. Thus intermodal containers became the basis of globalization. Containerization has changed the whole industry structure. Ships, trucks and railroad cars have been adapted to comply with container ISO norms. The whole logistics chain has changed. Jobs were switched from harbours to the countryside and standardization of all trade flows was established: Worldwide transportation costs have decreased dramatically.

Containerization, even more than float glass, is a novel combination of existing technology. But containerization only became successful due to global standards, with changes in infrastructure all over the world. It initiated billions of investments and at the same time made billions of investments obsolete!

12.3.3 Teflon: Serendipity at Work

Polytetrafluoroethylene (PTFE) became known under the brand name Teflon commercialized by DuPont Company. Roy Plunkett of Kinetic Chemicals in

New Jersey accidentally invented PTFE in 1938. While Plunkett was attempting to make a new refrigerant the tetrafluoroethylene gas in its pressure bottle stopped flowing and he found the bottle coated inside with a waxy white material, which was oddly slippery. The gas had turned into PTFE. It had become polymerized perfluoroethylene. The iron on the inside of the container had acted as a catalyst at high pressure. The process is still known as the Plunkett-process, after its inventor, or rather discoverer. Kinetic Chemicals patented the new fluorinated plastic in 1941 and registered the Teflon trademark in 1945.

At the beginning it seemed impossible to use the discovery commercially – manufacturing costs were too high. Nobody saw an application for the inert material. However, in 1943, the researchers of the Manhattan Project had no material to hold the highly reactive uranium hexafluoride. They remembered PTFE and used it to prevent corrosion in the manufacture of nuclear weapons.

Teflon made it from nuclear weapons to the kitchen (Roberts, 1989). The French engineer Marc Grégoire used PTFE on his fishing line to make disentangling easier. In 1954 his wife Colette urged him to try the material on her cooking pans. Marc Grégoire created the first pan coated with Teflon non-stick resin under the brand name of Tefal.

Due to its chemical inertia, Teflon is used when aggressive chemicals have to be contained or a low level of friction is needed. It is applied in sealing technology and in architecture for weather and UV-resistant windows. PTFE fulfils a similar function in Gore-Tex textiles. In medicine, PTFE is used for implants and for lenses in optics (Palucka, 2006). Even though Teflon is a radical innovation that was discovered by accident, it became accepted in many different areas due to continuous improvement.

12.3.4 smart: Planning the Unplannable

In contrast to "serendipity" and the accidental discovery of Teflon, the innovation of the compact car smart was a planned development project of Micro Compact Car (MCC) AG founded in 1994 in Biel, Switzerland. The idea originated from Nicolas G. Hayek, the founder of the Swatch group, today the biggest producer of Swiss watches. In order to realize his vision, he co-operated with the automotive manufacturer Mercedes-Benz AG. smart: *s* stands for Swatch, *m* for Mercedes and *art* for art. MCC aimed to manufacture a cheap and compact car for urban regions: the idea was that two smarts could fit crosswise into a parking lot. The realization started with a stringent and courageous concept: A compact car which never had existed before and only had two seats: "Reduced to the Max". Why buy a big car when we have on average only 1.1 passengers?

Not only the product, but also the processes were revolutionized: With the miniaturization of components and the recombination of Swatch- and Mercedes elements, smart became a "Lego type innovation". The modular sourcing concept was unique: smart, which is now a 100 % affiliated company of Daimler AG, started to produce in smartville, its factory in Hambach, France, and had only a handful system suppliers. Most system suppliers had some of their production located directly in smartville and delivered their modules directly to the smart assembly line. A smooth flow of the value chain was guaranteed. All partners had a 100 % single source long-term contract. smart centres ensured close customer contact based on franchise.

Smart is a unique combination of radical product and process innovation and consists of many small routine innovations which were planned as a green field approach and which were initially seen as radical. Everything was planned in detail, covered by a large amount of market research and backed at the beginning by two very experienced companies. Nevertheless, losses in the order of billions of Euros accrued and nobody knows whether a positive payback will ever be achieved.

12.4 Drivers of Radical and Routine Innovations

Radical and routine innovations differ in their consequences, but are driven by the same four major forces:
1. *Lower costs:* The original goal of automating the manufacturing process of float glass was to save costs. Henry Bessemer's first ideas did not succeed because the costs and resources were still too high. Customers could not be convinced. Sir Alastair Pilkington implemented the basic idea and demonstrated the industrial viability of the float glass process. Reducing costs is one of the most important drivers of most innovations. It is prevalent in many routine innovations: Even mature industries like the cement industry improve their productivity by 2 % every year without too much risk. Lower cost is always an advantage!
2. *Improved performance:* Containerization reduced the reloading times of ships significantly and lowered theft and transport damage. The transportation of supplies and spare parts in WW II was the original driver of this radical innovation. With the introduction of a radical new product, the intermodal container, the risk of this innovation could not directly be assessed. Neither harbours or railways nor trucking companies would have the courage to invest heavily. The risk of over-engineering always lurks behind performance improvement!
3. *New performance features:* The radical innovation of Teflon was developed by accident. With the new material PTFE, radical new performance features appeared which are very useful when dealing with aggressive chemicals or when the need for low friction arises. Teflon was first applied in the nuclear weapons industry and finally in pans, clothes and computer equipment. The application of and demand for Teflon of these new performance features could not be foreseen. Therefore risk and uncertainty in new technology is much higher than in cost reductions and performance improvements. Engineers had to find the applications for Teflon, this was the innovation!
4. *New competitive basis:* With smart, Nicolas G. Hayek promoted his vision of an integration of Swatch and a small car. A two-seated car for urban use was something new in a car world where all the cars get bigger and bigger. Despite

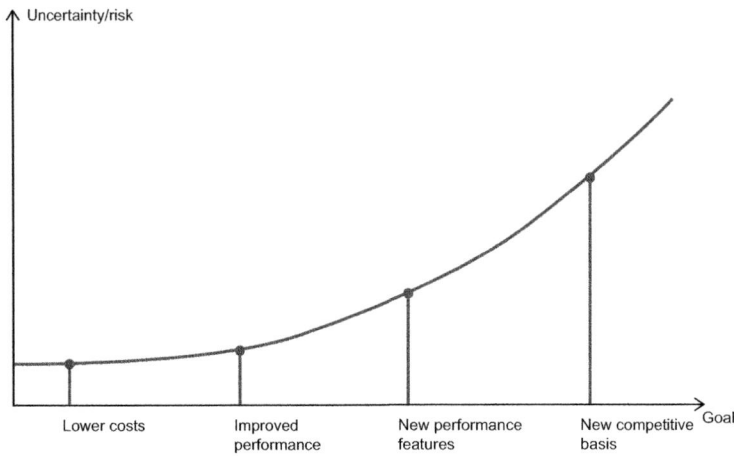

Fig. 12.2 Radical has many faces: somebody has to be surprised; speed makes the difference

the market niche, the introduction of smart was a high-risk innovation. Mercedes paid substantially: In 2008 the accumulated losses of MCC amounted to 8 billion Euros.

Although cost reductions and improved performance are usually drivers of routine innovation, radical innovations may also be based on such drivers, as seen with the float glass process or the containerization. In such cases the risks of radical changes can still be foreseen. However, the uncertainties increase when generating new features and changing complete systems. Nobody is able to predict the evolution of e-cars in the coming years, too much has to be changed by too many agents in the economy (Fig. 12.2).

For all radical innovations it is essential that somebody is surprised-users, customers, competitors or even the innovating company itself! Many radical innovations do not develop out of necessity, but out of abundance. The US experimented with CONEX before they were directly involved in the urgent transportation problems of WW II. Float glass, Teflon and smart were radical innovations that were discovered in times of abundance. Necessity may be the mother of some innovations, particularly for cost-cutting, and it is a weak justification for taking uncertainties and risks. This perception has changed the current innovation management of companies significantly.

In order to realize radical ideas, courageous actors were needed. Predecessors of Teflon, containers and flat glass had already existed and their ideas were only accepted due to the implementers, such as Pilkington, McLean and Colette Grégoire. Statistically many companies lack this pioneering spirit (Fig. 12.3): About 65 % of the produced development resources in companies are being invested in small routine product improvements that are cost-driven or due to customer complaints and competitive pressure. They lead to engineering changes to an existing product. About 30 % of the resources are being used to generate a

12.4 Drivers of Radical and Routine Innovations

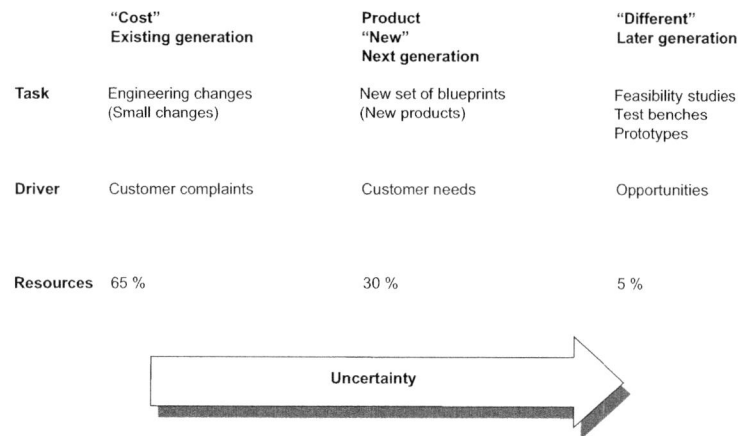

Fig. 12.3 Most companies do not spend enough on "thinking out of the box" for later generations

next-generation product. Only 5 % are invested in developing later generation products that could be radically different. In this case, feasibility studies, test benches and prototypes are necessary – a risk that many companies avoid because immediate urgency overcomes future needs.

In summary, there is anecdotal evidence:
1. Radical innovations can have the same drivers as routine innovations do, such as cost reduction and performance improvements.
2. Radical innovations can accumulate from many small routine innovations.
3. All radical innovations have one joint effect: surprise. This depends on the duration of the realization.
4. The more radical an innovation is, the more effort is needed to get acceptance, internally and externally.
5. Radical innovations are born rather in environments of abundance, less in ecologies of necessity and even less in cost reduction phases.
6. Radical innovations need all the skills of routine innovations.
7. Uncertainty grows with an increase of the radical nature of the innovation. Many companies avoid this risk.

Many radical innovations are hidden. You become aware of their revolutionary character only in hindsight. They are not planned. Containerization, mobile phones and SMS were caused by cultural revolutions that were driven by visions and values in a specific environment. These radical innovations arose without anyone understanding the overall effects. In contrast, the float glass process, smart cars and new medication are driven by cognition and grow parallel to routine innovations. August von Hayek distinguished between design (R&D), cultural evolution (learning) and Darwinian evolution (trial and error). Rational decisions lead to fast changes, but are only extrapolations. On the other extreme, Darwinian evolution may create radically new unexpected products through trial and error, but it takes time and it cannot be planned!

The Economist: Free exchange: The humble hero

Containers have been more important for globalisation than freer trade

© The Economist Newspaper Limited, London (May 18th 2013)

THE humble shipping container is a powerful antidote to economic pessimism and fears of slowing innovation. Although only a simple metal box, it has transformed global trade. In fact, new research suggests that the container has been more of a driver of globalisation than all trade agreements in the past 50 years taken together.

Containerisation is a testament to the power of process innovation. In the 1950s the world's ports still did business much as they had for centuries. When ships moored, hordes of longshoremen unloaded "break bulk" cargo crammed into the hold. They then squeezed outbound cargo in as efficiently as possible in a game of maritime Tetris. The process was expensive and slow; most ships spent much more time tied up than plying the seas. And theft was rampant: a dock worker was said to earn "$20 a day and all the Scotch you could carry home."

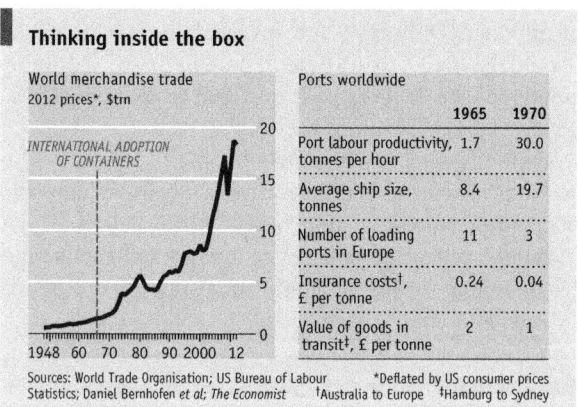

Containerisation changed everything. It was the brainchild of Malcom McLean, an American trucking magnate. He reckoned that big savings could be had by packing goods in uniform containers that could easily be moved between lorry and ship. When he tallied the costs from the inaugural journey of his first prototype container ship in 1956, he found that they came in at just $0.16 per tonne to load—compared with $5.83 per tonne for loose cargo on a standard ship. Containerisation quickly conquered the world: between 1966 and 1983 the share of countries with container ports rose from about 1% to nearly 90%, coinciding with a take-off in global trade (see chart).

The container's transformative power seems obvious, but it is "impossible to quantify", in the words of Marc Levinson, author of a history of "the box" (and a former journalist at *The Economist*). Indeed, containerisation could merely have been a response to tumbling tariffs. It coincided with radical reductions in global trade barriers, the result of European integration and the work of the General

Agreement on Tariffs and Trade (GATT), the predecessor of the World Trade Organisation (WTO).

Yet a new paper aims to separate one effect from the other. Zouheir El-Sahli, of Lund University, and Daniel Bernhofen and Richard Kneller, of the University of Nottingham, looked at 157 countries from 1962 to 1990. They created a set of variables which "switch on" when a country or pair of trading partners starts using containers via ship or rail (landlocked economies, such as Austria, often joined the container age by moving containers via rail to ports in neighbouring countries, such as Hamburg in Germany). The researchers then estimated the effect of these variables on trade.

The results are striking. In a set of 22 industrialised countries containerisation explains a 320% rise in bilateral trade over the first five years after adoption and 790% over 20 years. By comparison, a bilateral free-trade agreement raises trade by 45% over 20 years and GATT membership adds 285%.

To tackle the sticky question of what is causing what, the authors check whether their variables can predict trade flows in years before container shipping is actually adopted. (If the fact that a country eventually adopts containers predicts growth in its trade in years before that adoption actually occurred, that would be evidence that the "container" jump in trade was actually down to some other pre-existing trend.) But they do not, the authors say, providing strong evidence that containerisation caused the estimated surge in trade.

What explains the outsize effect of containers? Reduced costs alone cannot. Though containers brought some early savings, shipping rates did not drop very much after their introduction. In a 2007 paper David Hummels, an economist at Purdue University, found that ocean-shipping charges varied little from 1952 to 1970—and then rose with the cost of oil.

Put them in a container

More important than costs are knock-on effects on efficiency. In 1965 dock labour could move only 1.7 tonnes per hour onto a cargo ship; five years later a container crew could load 30 tonnes per hour (see table). This allowed freight lines to use bigger ships and still slash the time spent in port. The journey time from door to door fell by half and became more consistent. The container also upended a rigid labour force. Falling labour demand reduced dockworkers' bargaining power and cut the number of strikes. And because containers could be packed and sealed at the factory, losses to theft (and insurance rates) plummeted.

Over time all this reshaped global trade. Ports became bigger and their number smaller. More types of goods could be traded economically. Speed and reliability of shipping enabled just-in-time production, which in turn allowed firms to grow leaner and more responsive to markets as even distant suppliers could now provide wares quickly and on schedule. International supply chains also grew more intricate and inclusive. This helped accelerate industrialisation in emerging economies such as China, according to Richard Baldwin, an economist at the Graduate Institute of Geneva. Trade links enabled developing economies simply to join existing supply chains rather than build an entire industry from the ground up. But for those connections, the Chinese miracle might have been much less miraculous.

Not only has the container been more important than past trade negotiations—its lessons ought also to focus minds at future talks. When governments meet at the WTO's December conference in Bali they should make a special effort in what is called "trade facilitation"—efforts to boost efficiency at customs through regulatory harmonisation and better infrastructure. By some estimates, a 50% improvement in these areas could mean benefits as big as the elimination of all remaining tariffs. This would not be a glamorous outcome, but the big ones seldom are.

Sources

"Estimating the Effects of the Container Revolution on World Trade", by Daniel Bernhofen, Zouheir El-Sahli and Richard Kneller, Lund University, Working Paper 2013:4, February 2013

"Transportation Costs and International Trade in the Second Era of Globalisation", by David Hummels, Journal of Economic Perspectives, 21(3): 131-154, 2007

"Trade And Industrialisation After Globalisation's 2nd Unbundling: How Building And Joining A Supply Chain Are Different And Why It Matters", by Richard Baldwin, NBER Working Paper 17716, December 2011

Economist.com/blogs/freeexchange

References

Kim, W., & Mauborgne, R. (2005). How to create uncontested market space and make competition irrelevant. In C. Kim & R. Mauborgne (Eds.), *Blue ocean strategy: How to create uncontested market space and make competition irrelevant*. Boston, MA: Harvard Business Press.

Kuhn, T. (1965). *The structure of scientific revolutions*. Chicago, IL: University of Chicago Press.

Leifer, R., McDermott, C. M., & O'Connor, G. (2000). How mature companies can outsmart upstarts. In R. Leigh (Ed.), *Radical innovation: How mature companies can outsmart upstarts*. Cambridge, MA: Harvard Business Press.

Levinson, M. (2010). *The box: How the shipping container made the world smaller and the world economy bigger*. Princeton, NJ: Princeton University Press.

Miller, L. (2006). An analysis of sea shipping as global and regional industry. *Senior Honors Projects*, Paper 4.

Palucka, T. (2006). The slipperiest solid substance on earth. *MRS Bulletin, 31*, 421.

Roberts, R. M. (1989). *Serendipity: Accidental discoveries in science*. Toronto, Ontario: Wiley.

Rushton, A., Oxley, J., & Croucher, P. (2004). *The handbook of logistics and distribution management*. London, UK: Kogan Page.

Schumpeter, J., & Fels, R. (1939). *Business cycles* (1989th ed.). Cambridge, MA: Cambridge University Press.

Trend Towards Routine Innovation 13

> *Economic growth occurs whenever people take resources and rearrange them in ways that are more valuable. A useful metaphor...comes from the kitchen. To create valuable final products, we mix inexpensive ingredients together according to a recipe.... History teaches us that economic growth springs from better recipes, not just from more cooking.*
> Paul Michael Romer, American Economist, born 1955

Abstract

Companies are not tinkering around anymore or waiting for a radical innovation to evolve, but trying to routinize innovation. This trend is supported by the standardization of development processes, miniaturization and modularization of products, as well as a shift from competition to co-opetition.

Over the last few years, radical innovation was at the centre of scientific discussions. Due to numerous technical innovations, the decade of the 1980s was called the "Age of Discontinuity" (Drucker, 1986). For most companies today, innovation is a must in order to stay competitive. It is much easier to innovate today than it was 50 years ago. Companies are not waiting for a radical innovation to evolve, but opting for systematic innovation. Thus innovation as a tool is being pushed very far.

13.1 Standardization

Since the 1980s process management has been a central element of many quality-driven initiatives such as Total Quality Management, Business Process Engineering, Six Sigma, ISO and Lean Management (Benner & Tushman, 2003). Today every organization is urged to structure and continuously improve its processes. Process management is applied to all areas of management, even innovation. Some companies try to transfer the lean principles of the Toyota Production System to the development of products. A systematic design process starts with a customer need (Fig. 13.1). The need is analysed, transformed into a merit function and translated to a problem that can be simulated and represented by analytical models based on

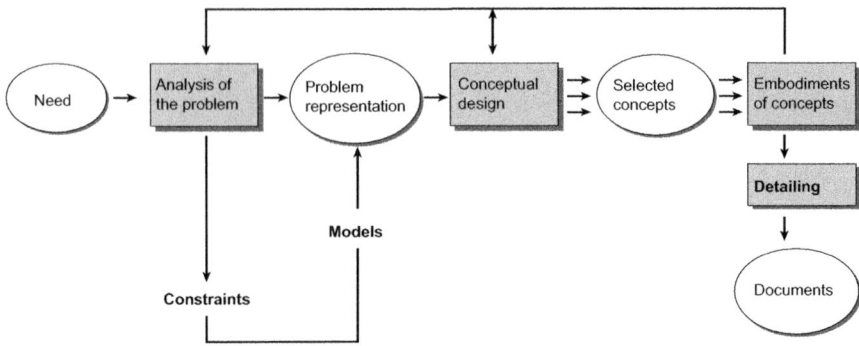

Fig. 13.1 The design process: from needs to documents (French, 1992)

natural sciences or even behavioural economics. Thus it can be solved with the help of computers. The solution to the problem is then conceptualized in a variety of alternative designs from which an optimal solution is selected and detailed in documents, such as blue prints and bills of material, ready for implementation.

Important prerequisites for process management are reproducibility and routine. Processes should not be individual or exceptional. We may only talk about processes if a certain structure is visible. Many argue that in today's R&D departments routine should not be applied because the environment is changing and new challenges frequently arise. But small changes with high predictability are good enough for most companies. A certain level of standardization is necessary to survive in many industries. Most industries are mature and radical changes could destroy big investments not yet written off. Thus many small changes at a high frequency make economic sense. In business one has only to be better than the competition. That is why in many fields it is better to build up the flexibility to respond to big changes quickly rather than promote radical innovation with its unknown unknowns! If an approved and tested process works successfully, it should be used as a standard.

A generic technical innovation process starts with

- An analysis of the environment (materials, manufacturing processes, existing solutions) to ascertain the most important constraints;
- An analysis of the conditions to be fulfilled in order to satisfy the customers (quantification of customer needs) to put together a specification list;
- A first technical concept for a solution to focus the minds in the R&D department.

In optics, routine innovation has been around for a long time. Standard everyday optics, like average camera lenses for example, relies on one physical law – the law of refraction – while the technologies of glass-making and lens manufacturing have been continuously improved over 200 years. Making glass which meets optical tolerance means reliable, low variation processes. Homogeneous glass makes it possible to manufacture lenses to extremely tight tolerances: A lens in a camera is polished to a deviation from a perfect sphere of less than 0.1 μm, which is 0.1

millionths of a meter. Thus in optics we have a precise physical model, material parameters with low variation and manufacturing capable of reaching reasonable tolerances. Companies such as Olympus, Nikon, Canon and Zeiss are able to calculate the next innovation. Not the generic concepts, but the specific lenses needed.

Seven steps of design lead from the clients' requirements to blue prints and a bill of material:
1. Classifying client's requirements
2. Identifying environment
3. Analysing, modelling
4. Identifying constraints
5. Testing, evaluating
6. Refining, optimizing
7. Documenting

The concept is fed into a computer which optimizes the solution by varying the parameters within a given set of boundaries. The final result is adapted to the manufacturing capabilities and the available material. This adaptation means setting tolerances for manufacturing that are driven by customer sensibilities, manufacturing capabilities and cost. Tolerances can be simulated only to a limited extent: We still have to rely on experience and trial and error. There are too many combinations to be calculated on a computer. If we have 10 parameters in our design that influence each other and we want to test 10 different values for every parameter we have to calculate 10 to the power 10 situations, which takes too long even on a big computer. Thus Monte-Carlo methods are applied. One simulates only a random sequence of parameters.

The whole design process goes back and forth between the designer and the computer. The designer develops a profound understanding of the system and is thus able to guide search algorithms setting priorities, freezing and unfreezing the constraints and parameters. The final design is tested on the computer for sensitivities to manufacturing and material tolerances and handed over to manufacturing in the form of drawings and a bill of materials.

As soon as we have a modular design with self-contained modules and weak interactions between the modules, design becomes much easier and computer times go down by orders of magnitude. We are able to create a new system much like a Lego figure. Modularization has made huge progress thanks mainly to miniaturization.

13.2 Miniaturization

According to Abernathy and Utterback, the technology life cycle can be divided into two phases (Abernathy & Utterback, 1978): In the early life of a technology, innovations focus on functional improvements of the product. After the emergence of a dominant design of the product, cost reductions and process innovations become more important.

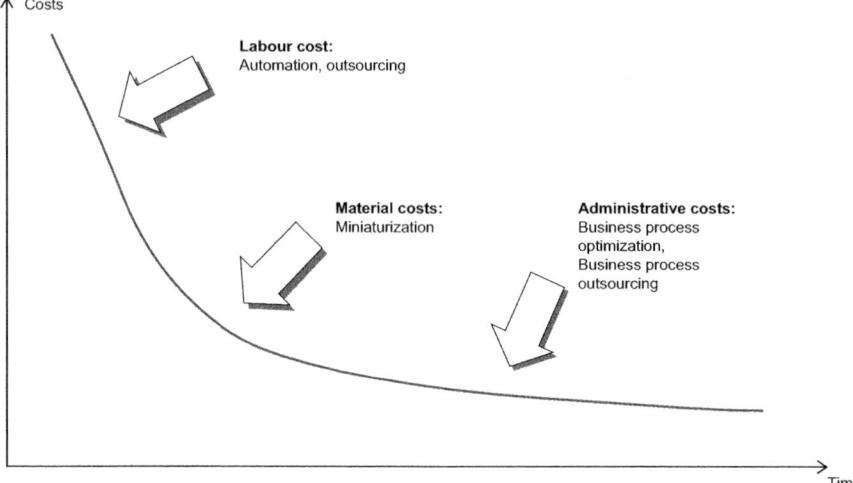

Fig. 13.2 Process innovations due to cost reduction follow a typical pattern (Boutellier & Rohner, 2007)

Process innovations that reduce production costs follow a typical pattern (Fig. 13.2):

Initially the biggest potential in savings is reached through a reduction of labour costs. This can be achieved through automation or outsourcing to achieve economies of scale or at least low labour cost. Miniaturizing components and parts decreases material costs. Finally, administrative costs can be lowered with business process optimizations. Automation and standardization reduce flexibility. Innovation has to conform to existing capabilities and becomes more and more routine. This does not imply lower innovation performance. In contrast, successful miniaturization can lead to new applications or can form the basis for further product innovations. One example is the ABS system. Besides higher flexibility to follow ever-shorter regulation cycles, the automotive industry required ABS systems with better reliability and lower weight. Reliability problems forced Bendix, a competitor of Bosch, to take their ABS system off the market 2 years after market introduction in 1971. In contrast, the improvements due to miniaturization of Bosch's ABS 2-System from 1978 to the ABS 8.1 in 2007 are enormous: The weight was reduced from 6.3 to 1.6 kg and the number of parts decreased from 140 to 16 (Boutellier & Rohner, 2007).

Something similar happened with the spark generators of Agie-Charmilles that merged with Georg Fischer Corporation in 2007. Over 30 years, its weight was reduced by a factor of 8 and finally only weighed 2 kg.

With miniaturization, system performance increases and modules, parts and components can be combined more easily. But there is no free lunch: Miniaturization needs tighter tolerances for material and production processes as well. Tighter tolerances can be achieved with better-trained staff or even through automation alone. Low labour cost loses its importance!

In software we do not have miniaturization, software statements do not get shorter and the total amount of software in daily-use goods is on a sharp rise: "The Internet of things" forecasts ubiquitous computing, tiny computers in a sea of data. We have nevertheless an analogue trend: Computers become increasingly powerful. Software may be set up in modules, standard sub-routines as well, even if this needs far more computing power than an integrated approach from scratch for every application. Many programmers see this as the productivity revolution in software development. We have known for many years that the most efficient programmers reuse and reuse code they have programmed in the past, whereas novices always start from scratch.

13.3 Modularity and Co-opetition

For many industries, such as software, aerospace and machining, the development of highly productive and complex systems is the basis for innovations but also one of the most difficult problems in engineering and management (Picot & Baumann, 2007). One technique to reduce complexity is modularity. A system is modular if it is broken down into subsystems, modules that are hopefully already on the market and used for many different products. Mutual dependencies of these subsystems should be reduced to a minimum or to a few defined interfaces, if possible standard ones. The key of modularity is that all modules are independent and convertible (Baldwin & Clark, 1997) or that there is only information flow between the different parts. In traditional technology like a locomotive at the beginning of the twentieth century – we have the "flow of mechanical power" with all its gears, cogs and levers. In today's locomotives we have electrical power.

The "power of modularity" is felt in R&D as well. It eases and optimizes the development tasks and makes them more flexible, more autonomous (Picot & Baumann, 2007). Autonomous innovation means an independent development of the individual modules. Whenever different modules are connected by standardized interfaces, each module can be changed or improved autonomously without any negative effect on the rest of the system. Thus individual modules can be developed in parallel and in various locations. Design, development and testing processes can be shortened significantly.

Miniaturization and modularity with standard interfaces are two trends that have created new industry industry-structures. Whenever a product is split up into modules, R&D can be delegated to the suppliers of the modules, usually with an increased economy of scale: The system producer can concentrate its own resources on system design. The supplier can sell its module for other applications as well. An economy of R&D and manufacturing evolves (Fig. 13.3).

Since functions can be incorporated into modules and these modules continue to get smaller and smaller, they can be easily combined into products. Typical examples are mobile phones and cars. In the Lexus 300 hybrid batteries, electric

Fig. 13.3 We expect a wave of innovations: combination of modules, but also combination of suppliers?

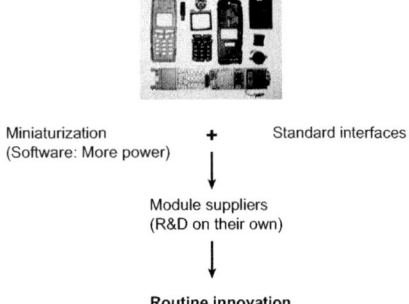

Miniaturization (Software: More power) + Standard interfaces
↓
Module suppliers (R&D on their own)
↓
Routine innovation

engines, petrol engines and control systems are so much smaller and lighter than 10 years ago and can thus be built into a car with astonishing features: It has standard dimensions, weighs some 2 t, goes from 0 to 100 km/h in less than 8 s and burns less than 8 l of petrol for 100 km in an urban environment.

However, as the product structure is decoupled and only standard interfaces exist, engineers have to communicate differently with one another. Integrated work environments can ease the communication network (Boutellier, Ullman, Schreiber, & Naef, 2008).

The combination of modules will lead to a wave of new innovations – modular innovations. Modules can be combined to form new innovations, "mix and match". We may call them "Lego-type innovations". Current examples of such combinatorial innovations are Smartphones that combine the functionalities of MP3, camera, organizer and phone etc. in one mobile phone. The principle is not new: Even Gutenberg used the concept of modularity for his invention of the printing press. He combined the principles of wine pressing, metal casting and cloth printing and adapted them to his printing requirements.

With modules it is easier to experiment in parallel, to follow a "best of breed approach" and then to choose the best design variant. Modular architecture means that the design process of complex systems can be conducted at a greater level of independence and with a greater flexibility. Modularity increases the routine characteristics of the innovation process and may increase R&D speed by factors.

Modularity has a significant impact on work sharing and competition within an industry. The development of new products is today characterized by specialization and by the integration of specific functions. Co-operation between suppliers and OEMs is one central element of staying competitive. This leads to a combination of co-operation and competition, called co-opetition. Co-opetition occurs when companies work together in one area of their business where they do not compete directly but where they can share costs. Peugeot is becoming more and more an engine producer for some of its competitors on the car market. Whatever is below the "line of visibility" in cars is seen less and less by OEMs to be important for differentiation. Why not reduce cost for everybody in the industry? The real competitor may be public transportation! Co-opetition favours the routinization of innovations.

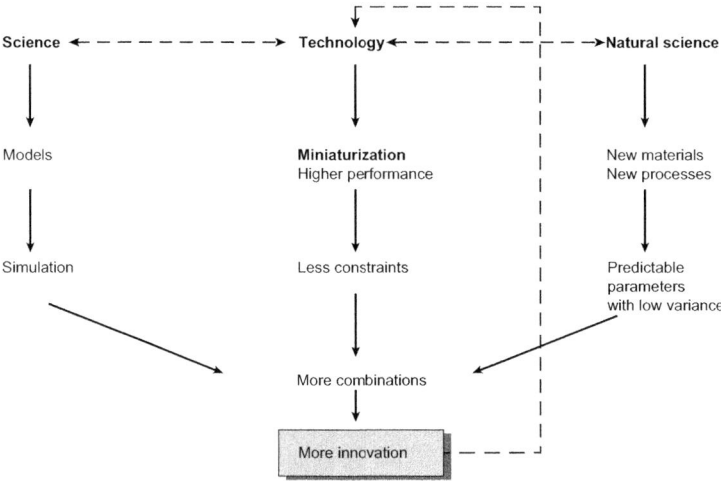

Fig. 13.4 Innovation becomes a routine: models and miniaturization

13.4 Models and Simulation

Science provides the basic structure of our models used to calculate new products. It leads to better materials and miniaturizes modules that perform a specific function.

Understanding a real object is one thing, teaching a computer to simulate a real object is much more demanding – and for companies, it is far more rewarding: Reality as described by a mathematical model can be optimized easily; innovation can be done on the computer.

Science improves models, technology miniaturizes with equal or higher performance, and new materials and processes result. Uncertainties and risks within innovation projects can be reduced which will lead to more routine innovations in the future (Fig. 13.4).

The Economist: Smart systems: Augmented business

Smart systems will disrupt lots of industries, and perhaps the entire economy
© The Economist Newspaper Limited, London (Nov 4th 2010)

CALL it the democratisation of sensors. Pachube (pronounced "patch-bay"), a start-up based in London, offers a service that lets anybody make sensor data available to anyone else so they can use them to build smart services. One tinkerer has Pachube's computers control the fan in his office, guided by temperature readings uploaded from a thermometer on his desk.

Such experiments are free, but those who develop more serious applications and do not want them to be available to anyone else have to pay. Usman Haque, Pachube's boss, hopes that more and more firms will do so as sensors multiply.

Pachube's business model is one of the more interesting attempts to make money from the convergence of the physical and the digital worlds, but there are plenty of other firms trying to cash in on smart systems. Many will fail, but those that succeed will disrupt more than one industry and perhaps the economy as a whole.

But what is most exciting about smart systems is the plethora of new services and business models that they will make possible. "The internet of things will allow for an explosion in the diversity of business models," says Roger Roberts, a principal at McKinsey and one of the authors of a recent study on the industry.

It is not just utilities that will benefit from smart systems but other sectors too. The chemical industry, for instance, has already installed legions of sensors and actuators to increase its efficiency. Others are just starting. In the paper industry, according to the McKinsey study, one company achieved a 5% increase in its production by automatically adjusting the shape and intensity of the flames that heat the kilns for the lime used to coat paper. FoodLogiQ, a start-up, allows food suppliers to tag and trace their wares all along the supply chain—and consumers to check where they come from. Sparked, another start-up, implants sensors in the ears of cattle, which lets farmers monitor their health, track their movements and find out a lot of other things about their animals that they never realised they wanted to know. On average, each cow generates about 200 megabytes of information a year.

Thanks to detailed digital maps, maintaining such things as roads and equipment will also become much more efficient. Asfinag, an Austrian firm, used aeroplanes equipped with special cameras to map the country's highways. Its employees can now fly over them digitally and even see what is underground. San Francisco's Public Utilities Commission knows the exact co-ordinates of every waste-water pump, its maintenance history and the likelihood of it failing. "Firms can now send out maintenance crews before things actually break," says Steve Mills, who heads IBM's software business. "Making the old stuff run better will be the most important benefit of such systems in the short run."

Moreover, smart systems make new forms of outsourcing possible, some of it to unexpected places. Pacific Control is not exactly a household name, but the company, based in Dubai, claims (with some justice) that it is the "world leader in automation solutions". Its global command centre remotely monitors buildings, airports and hotels, keeping an eye on such things as energy use, security and equipment. For the moment most of the firm's customers are in Dubai itself, but it should find more than a few abroad.

Significantly, once devices are connected and their use can be metered, there is no longer any need to buy them. Already some makers of expensive and complex equipment no longer sell their wares but charge for their use. Rolls-Royce, for instance, which makes pricey aircraft engines, rents them out to airlines, billing them for the time that they run. Makers of blood-testing equipment have taken to charging only if the device actually produces usable data. And Joy Mining Machinery, a maker of mining equipment, charges for support by the tonne.

Some firms are using metering in innovative ways. Zipcar and other car-sharing firms, for instance, put wireless devices with sensors into their vehicles so that customers can hire them by the hour. And insurance firms, among them Progressive in America and Coverbox in Britain, ask customers to install equipment in their cars that can measure for how long, how fast and even where a car is driven. Premiums can then be based on individual drivers' behaviour rather than on such proxies as age and sex.

Michael Chui, a co-author of McKinsey's report on the internet of things, says that such applications will allow companies to "have a much more dynamic interaction with customers". In Japan, for instance, vending machines can recognise a customer's age and sex and change the message they display accordingly.

The more data that firms collect in their core business, the more they are able to offer new types of services. By continuously assessing the performance of its jet engines around the world, Rolls-Royce is able to predict when engines become more likely to fail so that customers can schedule engine changes. Heidelberger Druckmaschinen, whose huge presses come with more than 1,000 sensors, has started offering services based on the data it collects, including a website that allows customers to compare their productivity with others. "Many companies will suddenly discover that their main business is data," says Paul Saffo, a Silicon Valley technology forecaster who wrote a widely noted essay on the impact of ubiquitous sensors back in 1997.

Hewlett-Packard is a prime candidate for such a Saffo moment—if its plans to scatter millions of sensors around the world come to fruition. It is doing this to increase demand for its hardware, but it also hopes to offer services based on networks of sensors. For instance, a few thousand of them would make it possible to assess the state of health of the Golden Gate bridge in San Francisco, says Stanley Williams, who leads the development of the sensors at HP. "Eventually", he predicts, "everything will become a service."

Apple, though it prides itself on its fancy hardware, is already well on its way towards transforming itself into a service and data business thanks to the success of its iPhone. When the computer-maker launched the device in June 2007, it did not expect "apps", the applications that run on smartphones, to become such a big deal. The "App Store", where users can download these pieces of software, was launched a year after the first iPhone was shipped. But the App Store now sells more than 250,000 apps that have been downloaded over 6.5 billion times. And with its new platform for mobile advertising, called iAd, Apple has started to make money from all the data it collects.

Some firms will make a living based entirely on mining "data exhaust", the bits and bytes produced by other activities. One example is Google's PowerMeter, which not only lets users check their use of electricity online but gives Google access to lots of data to analyse and, not least, sell advertisements against.

Conventional services, too, can be metered. Those supplied by governments may not be the first to spring to mind but, as a study by the Dutch economics ministry asks, why not use sensors for taxing things like pollution? That might be uncontroversial, but analytics software could also be put to more manipulative uses by fine-tuning charges for public goods to get citizens to behave in certain ways.

Much of the innovation in this field may come not from incumbents but from newcomers, and it may happen fastest on such platforms as Pachube. In a way it is a cross of YouTube and Windows. What made the video-sharing site so popular was the way it converted all videos to a common format. Pachube is doing the same for data feeds from sensors. And like Microsoft's operating system for applications, it provides basic features for smart services, such as alerts, data storage and visualisation tools.

This spring at Where 2.0, a technology conference in Silicon Valley, the star of the show was Skyhook Wireless, a firm that offers geographical-location information as a service. It recently launched a new offering called SpotRank. Drawing on all the data it has collected in recent years from apps using its services, the firm can predict the density of people in specific urban areas—anywhere, any day and at any hour. "This will give us great insights into human behaviour," says Brady Forrest, the chairman of the conference.

Another hit at the conference was a service called Wikitude. Its "World Browser", a smartphone app, checks the device's location as well as the direction in which its camera is pointing and then overlays virtual sticky notes other users have left about things like local landmarks. So far Wikitude and similar services are mostly used as travel guides. But in principle users could collaboratively annotate the entire physical world—and even other people. TAT, another start-up, already experiments with something called "Augmented ID", which uses facial recognition to display information about a person shown on a smartphone's screen.

It is difficult to say what effect all this will have on business and the economy. But three trends stand out. First, since smart systems provide better information, they should lead to improved pricing and allocation of resources. Second, the integration of the virtual and the real will speed up the shift from physical goods to services that has been going on for some time. This also means that more and more things will be hired instead of bought. Third, economic value, having migrated from goods to services, will now increasingly move to data and the algorithms used to analyse them. In fact, data, and the knowledge extracted from them, may even be on their way to becoming a factor of production in their own right, just like land, labour and capital. That will make companies and governments increasingly protective of their data assets.

In short, we may be moving towards a "Weightless World", the title of a 1997 book by Diane Coyle about a future in which bytes are the only currency and the things that shape our lives have literally no weight. But for now, gravity has not quite been repealed yet.

References

Abernathy, W. J., & Utterback, J. M. (1978). Patterns of innovation in technology. *Technology Review, 80*(7), 40–47.

References

Baldwin, C., & Clark, K. (1997). Managing in an age of modularity. *Harvard Business Review, 75*(5), 84–93.

Benner, M., & Tushman, M. (2003). Exploitation, exploration and process management: The productivity dilemma revisited. *Academy of Management Review, 28*, 238–256.

Boutellier, R., & Rohner, N. (2007). Miniaturisierung führt zu einem Zerfall der Rohstoffpreise - eine Marktanalyse. In P. J. Gausemeier (Ed.), *3. Symposium für Vorausschau und Technologieplanung*. Heinz Nixdorf Institut, Paderborn, Germany.

Boutellier, R., Ullman, F., Schreiber, J., & Naef, R. (2008). Impact of office layout on communication in a science-driven business. *R&D Management, 38*(4), 372–391.

Drucker, P. (1986). The changed world economy. *Foreign Affairs, 64*(4), 768–791.

French, M. (1992). *Form, structure and mechanism*. London: MacMillan.

Picot, A., & Baumann, O. (2007). Modularität in der verteilten Entwicklung komplexer Systeme: Chancen, Grenzen, Implikationen. *Journal für Betriebswirtschaft, 57*(3–4), 221–246.

Design of Industrial Goods   14

> *When products are developed by rugby teams, simultaneous engineering, and concurrent engineering or integrated engineering - in place of the traditionally sequential relay-race approach - the importance of industrial design becomes obvious. It is no longer possible to think of industrial designers merely as people, who "wrap products in nice shapes und pretty colours", as one cynic described their traditional role. Instead, for the development process to be effective, they have to be equal members of the team almost right from the start of the process.*
>   Christopher Lorenz, Management Page Editor, Financial Times

Abstract

Industrial design conveys a message, differentiates the product or makes functionality easier to apply. It is driven by aesthetics and depends strongly on time and culture.

In the past years design has gained increased significance as a success factor of western industries. If, in the past, industrial design was seen as a marginal problem of product and packaging presentation, in the future it will move into core areas of strategic decisions. Enterprises which successfully distinguish themselves from their competitors by a strong design are, for example, Coca Cola, Braun, de Sede, but as well Rubis, a Swiss tweezer-manufacturer, an SME (Fig. 14.1).

A consistent company reputation is only obtained by a consistent design of its logos, products, documentations, fairs, architecture, and brand itself. The potential of design is not exhausted within the consumer goods range; it is equally important for capital goods and software as well. The famous Italian designer Pininfarina designed the Swiss railway locomotive 2000, the standard locomotive running on Swiss tracks for the last 10 years.

Design and aesthetics aim for differentiation, to create products which are different from competing products. However, contrasts which are too extreme can also be disturbing: If customers do not recognize the "language of the design" they have a problem, they feel excluded. Design is based strongly on simplicity and on repetition, on respect for the customer and on consistency, but also on direct

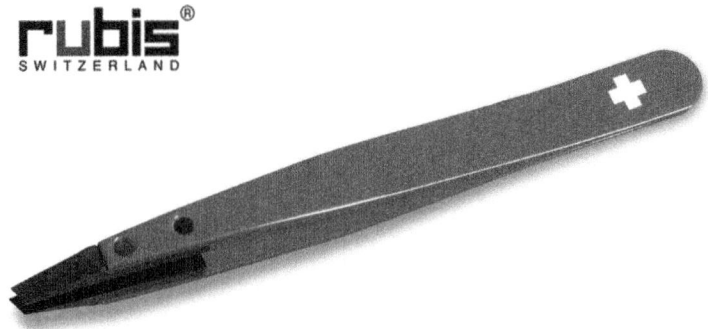

Fig. 14.1 Successful industrial design in Switzerland: Rubis

customer use: The K3 Yoghurt cup shows that some customers appreciate additional information. Design makes products desirable, besides factors such as usefulness and affordability (Fig. 14.2).

Design makes products unique. Coca-Cola supports its consistent reputation with the design of its bottles. Coca-Cola combines haptic with colour and its logo:

> ...a person could recognize it as a Coca-Cola bottle when feeling it in the dark, so shaped that, even if broken, a person could tell at a glance what it was...
> Leslie Novos, head marketing Europe Coca-Cola, 2002 Cerruti forum

14.1 Design Strategies

In principle, there are two generic design strategies: Inside-out and outside-in (Fig. 14.3). With the inside-out approach one wants to move functionality into the foreground and with outside-in it's the opposite, one wants to emphasize aesthetics.

14.1.1 Inside-Out Design

Both design approaches are justified: If a new technology is complex, customers are grateful if the design helps them to understand its functionalities on the spot, such as the design of a dashboard in a car. In the early phases of technology, inside-out design plays a fundamental role. It is important that how to use a machine is almost intuitive and does not entail reading long product descriptions. Most people do not read these descriptions, they just try. Geometry and activation mechanisms have to be as simple as possible. Everyone needs to understand how things work and therefore everything important has to be made visible. The user should be able to understand functionality without long explanations. On the other hand, everything

14.1 Design Strategies

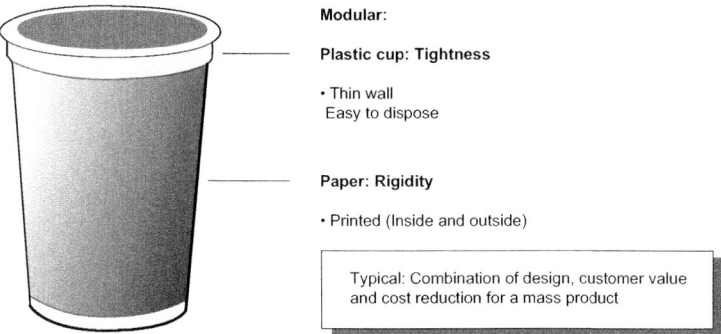

Fig. 14.2 Yoghurt packaging: adapted to logistics and added value for the customer

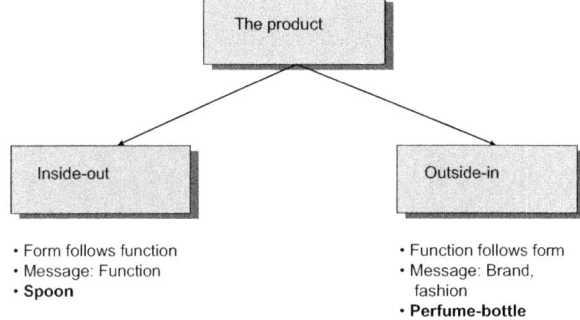

Fig. 14.3 Two generic strategies of industrial design: inside-out and outside-in

unimportant should be hidden from the line of sight, since it only alienates the customer.

A simple inside-out design is the well-known Swiss peeler of Zena whose design is "reduced to the max" (Fig. 14.4): You see it and you know how to use it.

The first peeler, called "Rex", with a horizontal blade, was invented in 1947 by the Swiss inventor Alfred Neweczerzal. In the same year it was patented. Rex is still being produced today. More than two million peelers are manufactured annually, of which 60 % are exported. In 2004 Rex was printed onto Swiss stamps as part of the series "Design classics of Switzerland".

14.1.2 Outside-In Design

In contrast, outside-in design focuses on shape and appearance. For this reason, functionality may suffer. Indeed, style and fashion are important and a degradation of functionality may be accepted. One ancient example is the first German edition of Vitruvius which appeared in 1564 in Basle (Fig. 14.5). It slowed reading down, which, as we know today, actually improves understanding! But that was not the intent of the editor: He wanted to impress customers with his printing capabilities

Fig. 14.4 "Reduced to the max": design without any explanations for use

Fig. 14.5 Outside-in design: function follows form

and to be as good as the well-known expensive and labour-intensive work carried out by monks in monasteries.

Among today's customers there is increasing acceptance of black boxes hidden behind simple man-machine-interfaces. Only a few people understand the details of an oil pump in their car, but they know they have to go to a garage whenever the red light "oil pump" comes on.

Outside-in design takes over as soon as the complexity of a technology has to be hidden from the eyes of the customer, or whenever it is more or less impossible to achieve differentiation through better technology.

14.1.3 Interaction Design

In software, design means design of interaction, not interface. As in many other design fields, interaction design also has an interest in creating a form which can be more or less elegant and more or less easy to maintain, but whose main focus is on behaviour. Interaction design is heavily focused on satisfying the needs and desires of the users, whereas other disciplines like software engineering have a heavy focus on designing for the technical stakeholders of a product. The program may have millions of lines of code, but the interaction with the user should remain simple and help the users achieve their goals. A simple way to achieve this may be the introduction of "personas" or "use cases": Representatives of user segments that have
- Specific goals
- Specific experiences with computers
- Specific skills
- Specific emotions

With the definition of a particular personality or individual, it is easier for software designers to find out what is desirable for the user: Either functional such as Novell Inc.'s software, or viable, such as Microsoft's software, or easy to use, such as Apple's software (Fig. 14.6).

It is important to keep the difference between customers and users in mind. Who makes the buying decision? If it is the customer, the one who pays our bill, brand may be more important than design: If you buy viable Microsoft, you go with the big majority. You cannot go wrong. But it may well be that your users will complain afterwards and you switch to Novell, the functional leader, that focuses on technology. Brand is important for the first buy, functionality may be more important for second buys.

User interaction designers have to remember that users are not average! Most interaction designs are made when the software has been developed and thus use an inside-out approach. They humiliate the users and make them feel stupid. Just think of your PC where you have to click on a field called START to stop your computer. Nobody complains about this anymore, only beginners are aware of it. Or take an average portable radio used at a fixed place in your kitchen. Whenever you start it, it tries to optimize, to find the best radio station. And you miss the news!

Fig. 14.6 Design = interaction design in software (not an interface!)

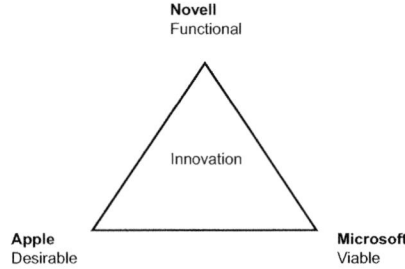

14.2 Design and Art

The importance of industrial design does not mean that artificial designs must arise. The functional surfaces invented by Heinz Isler, a Swiss construction engineer from ETH, show a balance and an elegance which emerged in a completely natural way (Fig. 14.7): Isler made no computer simulations as these did not exist in design engineering in his time. He made simple experiments. This is a procedure, which has led to success time and time again. With Christian Menn, Robert Maillart and Othmar Amman, Isler belongs to the most important Swiss design engineers of the twentieth century. Simple and natural processes inspired Isler. With frozen handkerchiefs, he found optimal surfaces and reduced the demand for design materials. Isler was especially popular for the construction of his 1,400 shell buildings that have a natural shape, minimal material consumption and low maintenance costs. However, they are difficult to build; their shapes are free-form surfaces (Ramm & Schunck, 1986).

To overcome this difficulty, Toyota is said to have reduced the shape of its cars to just a few mathematical curves, thus simplifying NC programming of its machine tools to speed up tooling and reduce costs.

Designs change with time, most visible in architecture. In 1905 the designers of the Hotel Löwen in Seelisberg, Switzerland, used an inside-out design with simple statics that supported the repetition of window modules; in 1985 the architect of the new Hotel Löwen used new materials for an outside-in design (Fig. 14.8). The old inside-out design was 80 % constant and had 20 % variations. Guests immediately felt at home as they saw what they expected. However, design was limited by constraints in materials and computation of statics. In contrast, the new outside-in design has low repetition and symmetry which is supported by unlimited design potentials, made possible by new technical processes and new materials. The visitors feel surprised and uncertain: More technology means more design alternatives, choices have to be made!

Fig. 14.7 A simple prototype with frozen water became the service area Mövenpick "Silberkugel" in Deitingen, Switzerland

Fig. 14.8 Inside-out design Pension Löwen in 1905 versus outside-in design Hotel Löwen 1985 at the same location

14.3 Design and Product Development

In the age of rapid prototyping, simultaneous engineering to shorten time to market and a smooth co-operation in interdisciplinary teams has increasingly become a must. Co-operation between R&D, production, marketing and design is the cracking point of the product development process. The designer is more and more a service provider and a catalyst complementary to the other disciplines. The designer has not only to understand the technical object; only through his early integration can cost and time savings be achieved.

The contribution of industrial design for product development lies on the time-critical path. An early and overall integration reduces time by avoiding change loops. However, this integration is not yet carried out everywhere in practice. Fast-moving consumer goods are best practice. Capital goods and equipment construction are catching up slowly. Enquiries show that approximately 80 % of product development costs are already specified in the initial stages of development which is a phase that is influenced strongly by design decisions. However, these decisions are mostly made by management or by representatives of marketing and R&D, unfortunately not by professional designers.

The cost of qualified design know-how amounts to less than 5 % of the total development budget for most technical goods. It is not possible to quantify the savings which result from an early inclusion of industrial design. However, a good designer understands the connections between design, materials and production processes and can therefore mediate between these areas.

At an early stage in the product development process, the industrial designer has to ask seven well-known questions of industrial design: How easily can one...
1. ...determine the functions of the product?
2. ...tell if the system is in a specific desired state?
3. ...tell what actions are possible?
4. ...determine the path from intention to action?
5. ...determine the path from system state to interpretation?
6. ...perform an action?
7. ...determine the state the system is in?

These questions are not easy to answer. Rules are needed. They vary from company to company and belong in many cases to the tacit knowledge of the organization.

Many designers work, therefore as Mozart did, with rather strict rules: They do not throw the dice, as suggested by Mozart, but always combine elements which are appealing or which make the users anticipate a specific message. Judgement and experience are necessary in order to weigh variation and constancy against each other. Design is communication and communication is simplification plus repetition. "To use the analogy of language, the composer will write in a language which the listener knows" (Pierce, 1980).

In Switzerland 90 % of industrial design is carried out by external consultants. Most of these design offices have between three and ten employees. Internal design departments are rare. Even large enterprises which are well known for their design, such as Swatch, obtain their design predominantly from external service providers. One design team takes responsibility for one product family or product series.

The main obstacles which make the overall integration of industrial design into the product development process difficult are:
- A lack of knowledge about the potential of industrial design. It is difficult to put a financial number on a successful design.
- Different points of view, technical languages and work methods of the functions involved
- Over-estimation of expenditures for qualified designers
- Varying performance evaluations and remuneration of the development team members

A design champion can facilitate integration. Their task is to be a mediator between the (external) designer and the in-house development team members. Usually they come from marketing or R&D and have a deep understanding of technology.

14.4 The Designer and Design Costs

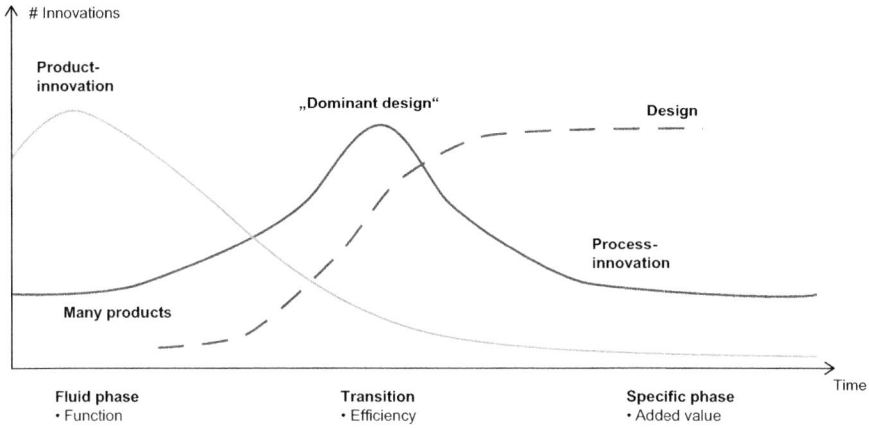

Fig. 14.9 Dominant design: from technical innovation to "design" (Utterback, 1996)

Design develops with products and processes: In the model of Utterback, design plays a substantial role only in the end phase (Fig. 14.9). However, many enterprises underestimate the potential of design before the dominant design emerges: For it is here that designers can mediate the basic message of a new technology and promote acceptance.

14.4 The Designer and Design Costs

Industrial designers should be involved in many different tasks within companies. They support shaping the vision, reputation and strategy of a company. From the strategy, a product program or "message" and a product concept for differentiation can be derived. This leads to product development where the designer has to trade-off aesthetics, costs and value added for the customer.

With their design knowledge, the industrial designer has a high impact on production costs. Thus a designer should also be integrated in the production processes and later in the product phase-outs in order to forward lessons learnt to new product development projects.

Thus designers have a very important position in companies. They are able to make or break them. Nevertheless, only a few design schools exist worldwide. The most famous ones are the Royal College of Art in London, the Pasadena College of Design in Los Angeles and the College for Creative Studies in Detroit. It is surprising that there are not many design schools in Italy or Japan![1]

Industrial design costs can vary greatly between products. Whereas the total design costs for production equipment and handheld medical instruments are low, its relative R&D cost is high, up to 30 % for handheld medical instruments, because

---

[1] The Economist, Dec 19th 2002.

in this case design serves functionality. The vacuum cleaner from Hoover needs more total design costs but is lower relative to its R&D costs. For cars and jumbo jets a lot of money is spent on design, but relative to total R&D costs, it is marginal.

The Economist: Italian fashion: Dropped stitches

A lack of skilled workers could kill one of Italy's manufacturing successes
© The Economist Newspaper Limited, London (Jun 22nd 2013)

JUNE is when the world's well-dressed look to Italy's big fashion shows to learn what to wear. This week the Pitti Immagine menswear show opened in Florence, followed a few days later by its Milanese counterpart, Milano Moda. But do not be fooled by the confident strutting of those hollow-cheeked clothes-horses: although Italian style still draws legions of buyers from the world over, beneath the catwalks the foundations of the country's flourishing clothing and accessories industries are being undermined.

"Within a generation the 'Made in Italy' label may be gone," frets Ermanno Scervino, a leading designer (one of his latest confections is pictured). Having survived the country's economic crisis relatively unscathed, Italian fashion, which generates a sizeable trade surplus, is contemplating its own extinction, simply because it is proving so hard to persuade young Italians to join the industry.

Behind the elegant clothing in Mr Scervino's boutiques in cities like London, Moscow and Tokyo is a range of artisanal skills—pattern-making, cutting, sewing, embroidery, knitting and the like—that turn his ideas into finished items. Hand-stitching dozens of diamond shapes in black chiffon for an evening dress, for example, calls for patience, good eyesight and a manual dexterity that comes only with much practice. Many workers at Mr Scervino's factory near Florence are well into middle age, and picked up their skills at home or in one of the small dressmaking and shirtmaking businesses that used to be numerous in Italian cities. As these workers head towards retirement it is unclear who will take their place.

What keeps Mr Scervino awake at night also troubles firms making the leather belts, bags and purses for which Florence is famous, and whose production gives work to around 12,000 people in and around the city. Gianfranco Lotti, who makes bags under his own name as well as for a global luxury brand, represents a lost Florentine tradition. Now approaching 70, he learned to make bags completely by hand through an apprenticeship that began when he was 14. Mr Lotti laments the shift from craftsmanship to machinery. Even so, skills are still needed, and the cost of teaching them is beyond the reach of many firms.

"The loss of know-how is dramatic," says Franco Baccani, the boss of B&G, which makes bags for Gucci, Cartier and other brands. Although it also relies partly on machines nowadays, B&G continues to rely on the human skills needed to select hides, cut and prepare them, and assemble and stitch the parts that make up bags. Shoemakers are in a similar fix. In the Marches region east of Florence, around 60% of the 700 production jobs at Tod's Group, a big maker of fancy footwear, are highly skilled; but despite offering good pay and conditions, and government-backed apprenticeships for raw recruits, the group is struggling to keep them filled.

With youth unemployment running at 35% in Italy and annual net pay for a young leather-cutter starting at around €18,000 ($24,000), fashion firms ought to have applicants beating down their doors. Like people in other rich countries, Italians tend to look down on manual work, however skilled, and families prefer to push their children towards careers in the professions and the public sector. The education system, at all levels, generally provides a poor preparation for working life. Italian universities are full of youngsters studying subjects in which they are not interested but which their parents think are good, regardless of the job prospects.

Around 86,000 jobs, mostly low-skilled, have been lost in Italy's textiles and clothing businesses since 2006, and many jobs in shoemaking have also gone, as work has moved to lower-cost countries. This may have given young Italians the impression that there is no longer any secure employment to be had in these industries.

However, the shortage of craftspeople is so widespread that some firms have taken to poaching from competitors, much as football clubs try to lure the best players from rival teams. So those with skills can be sure of finding work and commanding good pay. But the trade associations that represent makers of fashion goods have done a poor job of getting this message across. Only now are some industry bodies, and some of the larger firms, like Tod's, making an effort to promote careers in the business.

Some firms have looked abroad for skilled sewers and knitters. Mr Scervino has brought in knitwear specialists from Bosnia and Moldova, for example. But these, again, are typically middle-aged workers who will need replacing before long. Other Italian fashion firms have caused controversy by sending some sewing work abroad, bringing the pieces back to add the final stitches before slapping a "Made in Italy" label on them. To the more traditional firms, that risks undermining the label's cachet. But unless they can persuade more young Italians to come and work for them, it may be the only way to stay in business.

References

Pierce, J. (1980). *An introduction to information theory*. New York, NY: Dover.
Ramm, E., & Schunck, E. (1986). *Heinz Isler. Schalen*. Stuttgart, Baden-Württemberg: Krämer.
Utterback, J. M. (1996). *Mastering the dynamics of innovation*. Cambridge, MA: Harvard Business Press.

Protection of Innovation   15

> *Protection is not a principle. But an expedient.*
> Benjamin Disraeli (1804–1881), British politician

Abstract
Property rights are one of the cornerstones of Western civilization's success. Patents have been granted for a long time and copyrights as well. Both are being questioned in the age of the Internet and the age of the North–South divide.

The creative economy is growing fast: In the 1980s 40 % of the value of publicly traded companies was due to intangible assets; in 2005 they reached 75 %. Alan Greenspan, the chairman of America's Federal Reserve, said: "The economic product of the US has become predominantly conceptual."

In the US alone, USD 45 billion sales were generated through technology licensing in 2005, compared to USD 100 billion worldwide. Many technology-driven companies seek more and more patents and are building their business models around intellectual property while others just find ways of making money by opening up their innovation secrets and sharing them with others. Open-source software is one example of this.

Looking back in history we see that Italian and British Governments granted patents to individuals to reward them with a temporary monopoly, not for inventing, but for transferring foreign technology into their countries! In the past, patents were appropriated for the dissemination of technology and not for its protection. However, protecting the creative economy through an intellectual property system prevails today, especially in the US with their armada of lawyers: "Law is business". For the US, copyright products (i.e. books, music, TV programs etc.) are the primary export product. In 1997 the US economy manufactured products covered by copyright that accounted for more than USD 400 billion.

Today technology news travels fast. It is worthwhile to innovate only if we are able to protect our new ideas for at least a certain time.

Legally, protection of technology is built on four pillars (Fig. 15.1):
1. Patents: The aim is to protect inventions, technical ideas.
2. Copyrights: Used to protect artistic, musical or literary work.

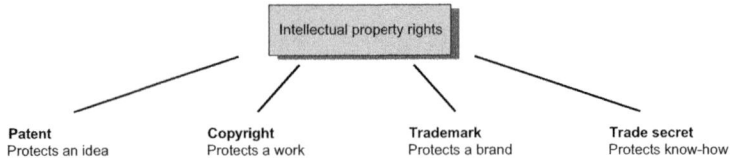

Fig. 15.1 Property rights are built on four major pillars, three protection rights and one contractual agreement

3. Trademarks: Used to protect brands.
4. "Trade secrets": Practices and knowledge that are kept confidential within a company.

According to Thomas Hobbes and John Locke, a property right follows from "natural rights" – a view which is still held in Europe. Both Hobbes and Locke are English philosophers who supported the theory of social contracts: An inventor produces new property that belongs to him. In contrast to this, the USA and Japan have a more utilitarian approach, which favours commercial exploitation over the creator's right. Japan has developed the practice of early licensing: As soon as a patent is published everybody can make use of it, but they have to pay a reasonable license fee to do so. If the patent is not appropriated later in the process, the licensees will get their money back. Governments strive to achieve a balance between protection which makes invention attractive for the inventor and dissemination of know-how to promote economic growth.

15.1 Patents

Patents and licences are important drivers of innovation as they give inventors an incentive to make their know-how publicly available and still enjoy some exclusivity. By fostering communication and knowledge exchange between R&D groups, patents are important tools for technology and innovation management. They have existed for several hundred years. Patent law tries to resolve the imbalance between expensive innovation and cheap re-engineering and copying. The patent gives its owner a temporary and geographically defined monopoly and excludes third parties from taking advantage of every invention. This promotes innovation, at least in the eyes of our governments. Some economists have second thoughts: Whether the benefits of today's patent system outweigh their costs is being questioned. The holder of a patent forbids people or companies from using and exploiting their specified process or product for some period of time. To put it another way, a patent is a "negative right" (Jaffe & Lerner, 2011).

Thus governments see patents less as property rights and more as drivers of innovation, as patented know-how is publicly available but gives the inventors some time to use it for differentiation. According to the Federal Court of Justice

(Bundesgerichtshof) on July 4, 1995 (BGHZ 130, 259) the foremost duty of patent law is:

> to encourage technical progress and to profitably promote the spirit of invention for industry

To secure patent protection for a technical idea, the idea has to pass four tests. The decision as to whether the idea passes the test or not is done by "persons skilled in the art", experts that have deep domain-specific know-how:

1. An invention must describe a technical idea, a technical solution for a technical problem.
2. The invention must be useful. It cannot be a mere theoretical phenomenon but must be realizable. This was the guarantee in the nineteenth century to make sure that scientific results could not be patented! Science was of no direct use! In the patent this is given through a potential realization.
3. An invention has to be novel, that is, it must be something that no one has disclosed before.
4. The invention should not be obvious. The larger the prior art, the stronger the patent. The average expert ("person skilled in the art") is assumed to possess knowledge of all existing technologies published in his field of expertise ("prior art"). In order to meet the patentability criterion of "inventiveness", an invention, for that expert, must not be obvious with regard to a combination of the prior art. Patent applications often refer to existing technology and show the improvements achieved.

If an invention passes all of the above tests, it needs formal registration. In most countries formal registration is conducted according to the "first-to-file" system that gives a patent for a specific invention to the first person that applies for it, regardless of when it was actually made. The US was unique in using the "first-to-invent" system where the date of invention, not filing, was the decisive factor. The US switched to a first-to-file system in March 2013.

15.1.1 Famous Patent Disputes

As patents exclude others from using an invention for 20 years in most countries, a patent grant can sometimes secure the existence of inventors. Thus it is not astonishing that several famous patent disputes over important technologies took place in the twentieth century: In 1901, G. Marconi had a 29-year lasting patent dispute with N. Tesla about who first transmitted radio signals. In 1903, Henry Ford started a patent dispute with the Association of Licensed Automobile Manufacturers on the first car assembly line. He won the dispute in 1911. The Wright Brothers patent application for a "Flying machine" was first rejected in 1903. With the help of a patent attorney, the US patent was finally granted in 1906. In the end, the brothers lost their dispute against the director of the Smithsonian Institution, Samuel Langley: In 1917, the federal government revoked their patent, which ended in a lawsuit that damaged the image of the Wright brothers significantly.

15.1.2 From Breadth to Specificity

Historically, patents were granted to people who did not make any discoveries at all. For example, in the seventeenth century Queen Elizabeth I granted 50 patents that gave their holders the exclusive right to manufacture and sell such basic materials as salt, wine or paper. Royals owned a large number of monopolies in important industries. By the eighteenth century two standards on patenting emerged in Great Britain: (1) a patent should only be granted for new and important technical inventions and (2) the breadth of the patent should be proportional to the importance of the discovery. Today's patents have to be very specific to get approval from the authorities.

15.1.3 Patenting Worldwide

Today there are some 50 million patent publications worldwide in 65 languages. Two thirds of these publications are in English, German or French. Only a small portion is still valid. The others have never received formal approval or have expired. But among the existing patents there are more than 15 million inventions and every day, some 10,000 join the existing lot. In the US the number of patent applications doubled in the decade from 1990 to 2000 and is still growing.

The tide of patent applications started when the US Congress converted the patent office from an agency funded by tax revenues into a "profit centre". As a profit centre, the patent office made money more or less proportional to the number of approvals. From then on, software, financial methods, biotech and business processes could be patented as well. In order to judge the numbers correctly, one has to keep in mind that one US patent equals four Japanese patents. The Japanese try to split their patents into smaller pieces in order to receive better protection and employees try to collect the money companies grant their patent developers. Japanese patents used to only allow the protection of one single "idea".

Patents are one of the best sources of documented technology worldwide. About 80 % of technology information is only published in patents. The fact that patents are up to date is shown by an example of a car tyres patent. The invention was patented on July 9th, 1981. In 1983 an internal presentation about the technology was shown to the employees, in 1987 it was published in the TCS magazine and in 1989 it was presented at a car show (Ehrat, 1997).

Patents are classified according to the international key International Patent Classification (IPC) that has more than 65,000 classes and is regularly adapted to technical progress. Further international harmonization is currently in progress. With the IPC you can compare patents which stem from different sources. Specialists in patent offices do the classification, making it objective and therefore

15.1 Patents

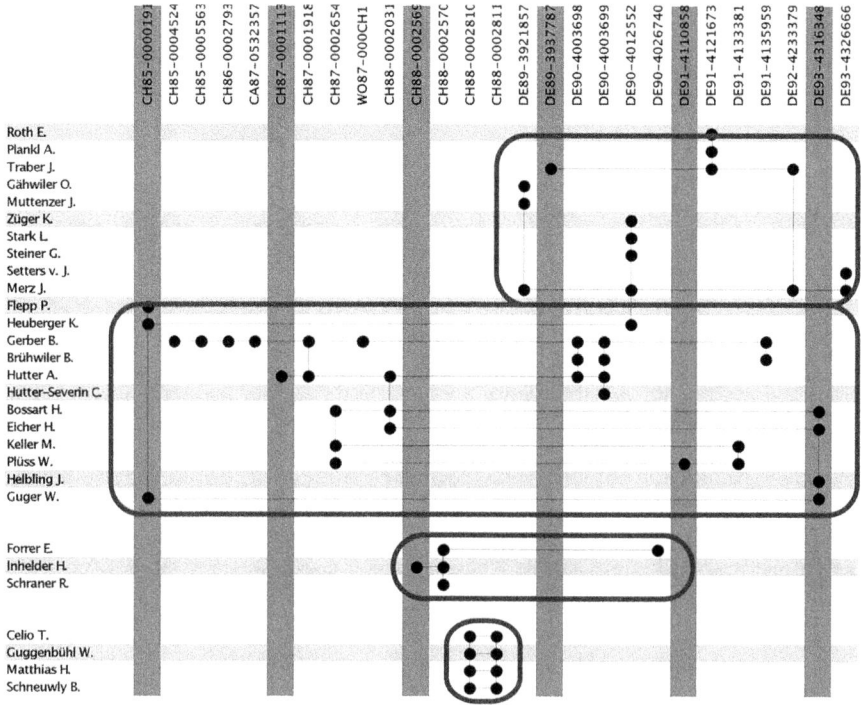

Fig. 15.2 Brain-mapping: patents show gate-keepers (Ehrat, 1997)

a good starting point for technology queries. Since computing power and search algorithms have improved by factors over the last 10 years, searching patents is a useful tool for identifying solutions to technical problems.

Patent analysis can show the attractiveness of new fields. Patent clusters may be a good indicator of background activities and the basic technological strategies of companies. Mapping patents with their originators (brain-mapping) might reveal gate-keepers of specific technologies, those key people who own specific knowledge about technologies and are at the centre of technology networks within their companies (Fig. 15.2).

After such an analysis we have the know–who needed: deriving know-how from the experts identified becomes relatively easy. For companies it is important to know gate-keepers in order to communicate both externally and internally: As engineers in general like to tell other people about their findings, they are prone to be spied on by competitors. A patent analysis shows the gate-keepers, the Internet shows where these specialists give talks and finally direct contacts. Showing interest helps to get the information one is looking for. Some companies train their employees to keep everything a secret as long as there is no patent protection.

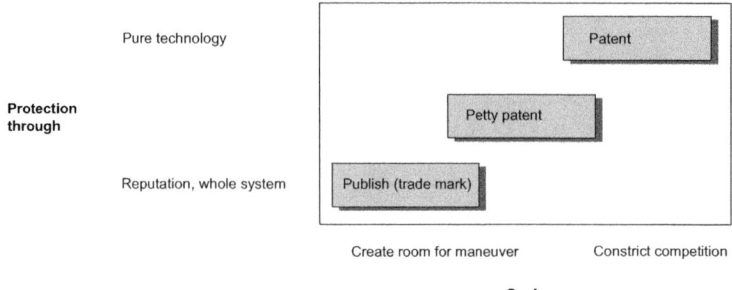

Fig. 15.3 Two fundamental strategies of protection: goals and enforceability are important

Patents are becoming more and more important for corporations. Patents make it possible for companies to stake out their lot and to secure room for manoeuvring. But once patented, your know-how becomes a public good. A patent may be by-passed. Once engineers know that there is a solution they become very innovative. If a company does not patent, it runs the risk that somebody else will file the patent and therefore it cannot use the technology anymore. Thus there is a second strategy: You do not patent, but you publish instead (Fig. 15.3).

You do not publish in Nature or Science but perhaps in a regional newspaper, hoping that competitors will not read the technical part of the text. By doing this you have evidence of prior use whenever a competitor claims priority. Thus companies only patent in important markets and in potential countries of production, but not everywhere. A well-known Swiss company published its innovations for many years in the Swiss bowling journal.

In some countries a new approach is emerging: Petty patents with less protection but also much less administration. This is reasonable wherever technology moves fast. There are very few electronic patents that keep their value for 20 years. Petty patents give a limited protection for 5–15 years. Germany provides 10 years.

15.1.4 Patenting in Europe

German and Swiss companies usually apply for European patents or international patents. In Switzerland you cannot patent software, mathematical methods and scientific theories (there are some subtle exceptions). Until 1981 software was not patentable in the US either. Likewise, an American court allowed business methods to be patented in 1998. This caused a high amount of applications such as the double click to confirm orders at Amazon. Europe has resisted patents on business methods and closely scrutinizes software patents.

A European patent is not valid in all EU countries. Three-decade long attempts to establish an EU-wide patent were blocked for language reasons and a lack of a centralized decision-making power. For example, the EU Directive COM of 1992/2002 proposed to view computer-implemented inventions as belonging to a

15.1 Patents

technical field which would make software patenting possible. In contrast, the European Parliament has issued a statement insisting "...data processing is not regarded as a technological field... data-processing inventions are not viewed as inventions in the sense of patent".

As long as there is no agreement the proposed directive will not be issued and the power to decide fundamental patent issues will be shifted back to the patent offices and courts. As experience grows, a European patent for software can be expected, but this process may drag on for years or even decades.

Application for an international patent from a specific country is either possible in the country itself or with World Intellectual Propert Organization (WIPO). The application first undergoes a formal check and is then forwarded to the individual states it applied for. The check for materiality is carried out by the individual states and is also granted state by state. Laws are a national matter. It is possible to have a central preliminary check with regard to whether the novelty and inventiveness criteria are fulfilled. To keep the costs at a reasonable level it is common to submit the application but final application and payments for all Patent Cooperation Treaty (PCT)-member states only in the most important markets and the most important manufacturing places.

The checks for an EU patent are carried out in Munich. Applications are possible in Munich, The Hague and Berne, for Swiss companies. After the European patent has been granted the individual countries manage it from there.

15.1.5 Patent Protection

The effectiveness of patents has been questioned for a long time. There are industries where hardly any patents exist anymore. But times are changing. Up until a few years ago, the banking sector had no patents, but since the double click with the mouse was patented in the US, patents in banks have been growing at a double digit rate for years. Microsoft had no patents until 2000. Then they hired the former head of IBM licensing and the company is speeding up on patent applications. In 2011 Google bought Motorola for billions in order to gain access to Motorola's huge patent basis: If you have no patents of your own, you cannot deter your competition! Today we have a serious patent war going on between Apple, Samsung, Nokia and Microsoft in the field of mobiles. The best protection is still having your own solid base of patents, especially in the US, where lawsuits are so expensive that most patent quarrels end in a compromise outside the courtroom.

Whenever a patent reveals too much information and its enforcement is difficult and expensive, it is better to keep important information secret. This is the case in process industries such as the cement and the food industry. Google is very secretive: In practice, you cannot protect search algorithms by making them publicly available! We should keep in mind that the law in most countries enforces secrecy not only for current employees but for past employees as well.

15.1.6 Cross-Licensing

Another method to make room to manoeuvre and to avoid unproductive lawsuits is cross-licensing. Big companies like Philips or Siemens have the same patent power as big Japanese firms like Sony or Fujitsu. Thus they mutually agree to refrain from enforcing patents among a selected group of companies. This is not viewed as a cartel by the anti-trust authorities because it enhances innovation. Patent rights are given priority over anti-trust rights. The down-side of cross-licensing is that third parties are out of the business. The big players keep the small ones at bay. The same strategy is sometimes applied by companies which have the information to prove prior art: They do not tear down the patent but buy a licence from the patent-owner, keep their mouth shut and all the others out.

15.2 Copyrights and Trademarks

There are not only patents but also other laws that protect ownership, such as copyrights. Napster, for example, changed the world's view of copyrights with its new technology of MP3 and peer-to-peer filesharing. This new technology allowed users to share MP3 songs easily. Downloading songs became easy as no server was needed; the individual database could be directly accessed. The co-founders of Napster, Shawn and John Fanning and Sean Parker had the intention of setting up a search engine, a direct filesharing system and an Internet relay for chatting. Napster specialized exclusively in music with the MP3 standard and presented a user-friendly interface. However, its distribution led to massive copyright violations. In the same year, in 1999, Napster was sued for billions by the big record companies. Although the original service was shut down by a court order in 2000, the Napster brand survived after other companies purchased the company's assets. With Napster, the whole music industry has changed and it is losing its protection: Today in many countries downloading is allowed, but law forbids uploading.

What the Napster case shows is that copyright protection changes with copying technologies: For example the painting of the Mona Lisa had natural protection, as a copy would have been as expensive as the original in terms of colours and working hours. Whenever a copying technology improves the copyright has to catch up before pirates like Napster take over.

The copyright is always running behind technology, but once in place, lobbying tries to strengthen the protection in favour of the copyright holders. Today's 70 year protection is a result of such actions and is detrimental to innovation in some industries.

In contrast to patents, copyrights do not need a specific format. All literature, artwork and scientific articles are protected by copyright from the outset. A copyright protects the originator for at least the author's full lifetime and in most states for another 70 years after his or her death. No international copyright exists.

But copyrights have important exceptions: For example, scholars and libraries allow copying for private or small-scale use.

The trademark law was introduced in Great Britain in 1897 to prevent bad merchandise from Germany being imported in order to protect the stagnant economy in England. By marking the foreign products with a "Made in..." label, British society labelled foreign products as inferior to domestic products. But the term *Made in Germany* was soon associated with product reliability and became a mark of quality.

Trademarks also have to pass a number of formal tests:
1. A trademark has to differ from existing ones within the realm of the goods and services to be protected.
2. The first one to use a trademark has priority.
3. A trademark is not allowed to be misleading nor can it be public property.
4. A trademark has to be applied for and used in public to gain legal protection (some national exceptions).
5. The one who does not use the trademark loses the right (in general a five year period of grace).
6. The trademark owner has to watch the market himself for infringements, whereas authorities do not.

Trademarks are sometimes better than patents: They last forever, i.e. as long as they are used, and it is much easier to control infringement. But the development of trademarks as well as the constant vigilance with regard to infringements make brands expensive. If you accept that somebody else has been using your trademark for several years, the other company can keep it. And introducing a trademark also means checking very carefully whether it has already been used somewhere else. Thus companies apply for protection as soon as they have names with a certain customer appeal. Natural words sometimes have unfavourable meanings in other languages. Companies use artificial names like Novartis or Coperion, Unaxis, etc. to get rid of all these problems.

The enlargement of the EU, the deregulation of many businesses in most states and the growing international competition increases the importance of trademarks. Nestlé has some 30,000 patents but some 120,000 trademarks, 6,000 of them used currently as trademarks. Trademarks and brands are the ambassadors of reputation (Table 15.1).

You need decades to build up a global name but you can destroy it within months, as shown by Arthur Anderson in the ENRON case. It makes a difference whether you have a title from a top university or from your local university. The Internet is heading in the same direction. Whenever you cannot immediately judge the quality of something, brands move in and tell the customer what to expect!

Table 15.1 The 10 most valuable brands 2012 (www.interbrand.com)

Trademark	Value 2012	Increase from 2011 (%)
Coca-Cola	$77.9 million	+8
Apple	$76.6 million	+129
IBM	$75.5 million	+8
Google	$69.7 million	+26
Microsoft	$57.9 million	−2
General Electric	$43.8 million	+2
MacDonald's	$40.1 million	+13
Intel	$39.4 million	+12
Samsung	$32.9 million	+40
Toyota	$30.3 million	+9

15.3 The Protection Mix

Innovations are only valuable in financial terms if they can be protected for at least a certain length of time. Patents, copyright, trademarks and secrecy are some of the means to achieve the appropriate protection needed by a company: It is the reasonable mix of these rights, not the specific use of one of them that makes the difference (Coradi, Heinzen, & Boutellier, 2013). Apple uses all these rights at the beginning of the twenty-first century to protect its business. Apple's I-Pod is a well-known trademark that can be used to download music protected by copyright. The technology is well protected with many patents and Apple is restrictive, even secretive about its new products. Technology-driven companies do not only need a marketing mix; they need an IP-protection mix as well (Fig. 15.4)!

Google is different: You cannot protect search algorithms with patents or copyrights. Legally this is possible, but once published, it is easy to turn around specific software. Thus brand and secrecy become important. But Google has bought Motorola and has now thousands of patents. They provide no direct protection of search algorithms; they protect all the other technologies Google uses. They help Google in its expensive patent fights with Microsoft, Samsung, Apple and all the other big mobile manufacturers.

The Economist: Smart-phone lawsuits: The great patent battle

Nasty legal spats between tech giants may be here to stay

© The Economist Newspaper Limited, London (Oct 21st 2010)

HISTORY buffs still wax poetic about the brutal patent battles a century ago between the Wright brothers and Glenn Curtiss, another aviation pioneer. The current smart-phone patent war does not quite have the same romance, but it could be as important.

Hardly a week passes without a new case. Motorola sued Apple this month, having itself been sued by Microsoft a few days earlier. Since 2006 the number of

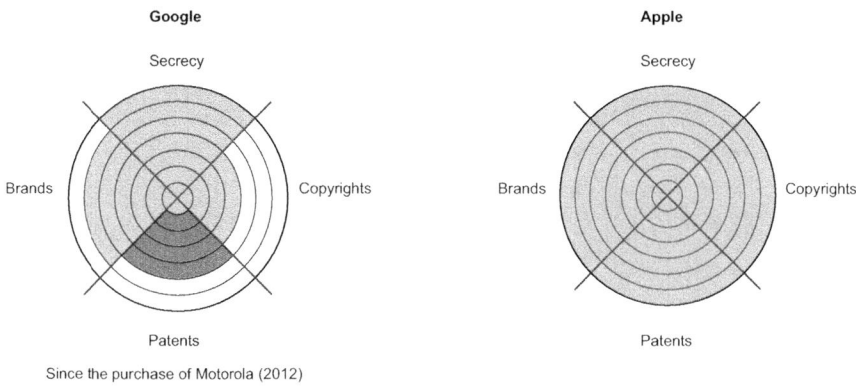

Fig. 15.4 IP protection mix of Google and Apple

mobile-phone-related patent complaints has increased by 20% annually, according to Lex Machina, a firm that keeps a database of intellectual-property spats in America.

Most suits were filed by patent owners who hail from another industry, such as Kodak (a firm from a bygone era that now makes printers), or by patent trolls (firms that buy patents not in order to make products, but to sue others for allegedly infringing them). But in recent months the makers of handsets and related software themselves have become much more litigious, reports Joshua Walker, the boss of Lex Machina.

This orgy for lawyers is partly a result of the explosion of the market for smart-phones. IDC, a market-research firm, expects that 270m smart-phones will be sold this year: 55% more than in 2009. "It has become worthwhile to defend one's intellectual property," says Richard Windsor of Nomura, an investment bank.

Yet there is more than this going on. Smart-phones are not just another type of handset, but fully-fledged computers, which come loaded with software and double as digital cameras and portable entertainment centres. They combine technologies from different industries, most of them patented. Given such complexity, sorting out who owns what requires time and a phalanx of lawyers.

The convergence of different industries has also led to a culture clash. When it comes to intellectual property, mobile-phone firms have mostly operated like a club. They jointly develop new technical standards: for example, for a new generation of wireless networks. They then license or swap the patents "essential" to this standard under "fair and reasonable" conditions.

Of apps and scraps

Not being used to such a collectivist set-up, Apple refused to pay up, which triggered the first big legal skirmish over smart-phones. A year ago Nokia lobbed a lawsuit at Apple, alleging that its American rival's iPhone infringes on a number of its "essential patents". A couple of months later, Apple returned the favour, alleging that Nokia had copied some iPhone features. Since then both sides have upped the ante by filing additional complaints.

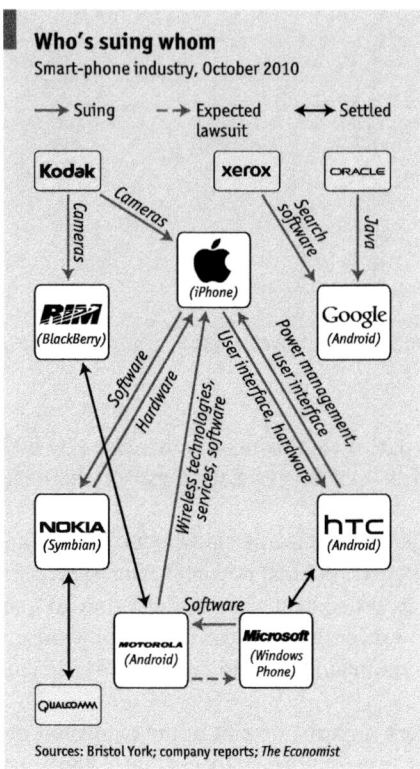

Lending ferocity to this legal firefight is the fact that competition in the smart-phone market is not merely about individual products, but entire platforms and operating systems (see chart). These are the infrastructures that allow other firms to develop applications, or "apps", for these devices. Should any one firm gain an important lead, it might dominate the industry for decades—just as Microsoft has dominated the market for personal-computer (PC) software.

Yet there is a difference between the smart-phone war and the earlier one over PCs. There is a new type of player: firms with open-source platforms. Google, for instance, which makes its money from advertisements, does not charge for Android (its operating system for smart-phones) and lets others modify the software. This makes life hard for vendors of proprietary platforms, such as Apple and Microsoft.

This explains another set of closely watched lawsuits. In March Apple sued HTC, a Taiwanese maker of smart-phones, which responded in kind a couple of months later. Earlier this month it was Microsoft's turn to file a lawsuit against Motorola, which is likely to retaliate soon. And Oracle, a maker of business software, has sued Google directly, albeit over a different issue, concerning how Android uses Java, a programming language.

Some expect Apple and Microsoft to sue Google. Yet this is unlikely, because the online giant will be hard to pin down. Google does not earn any money with Android, which makes it difficult to calculate any potential damage awards and patent royalties.

The frenzy of smart-phone litigation could last for years. The combatants have deep pockets and much to lose. Google is bounding ahead: Android now runs 32% of smart-phones sold in America, whereas Apple's iPhone has only 25% of the market and the BlackBerry has 26%, according to Nielsen, another market research firm.

Similarly, Nokia, which has already lost ground to Apple, cannot afford to see its intellectual property devalued. Should Nokia prevail, Apple may be slapped with $1 billion or so in licensing fees, estimates Christopher White of Bristol York, an investment bank.

If the market for computing tablets (which are essentially oversized smart-phones) takes off as expected, there will be even more lawsuits. Apple is set to sell some 15m iPads by the end of the year. Gartner, a market-research firm, predicts that the overall tablet market will reach nearly 55m units in 2011.

Eventually, even lawsuits must come to an end. How much harm they will cause remains to be seen. If Apple wins against HTC, that would be bad news for upstart handset firms. Until a few years ago, HTC only made devices for others, but now it has become a brand of its own.

Litigation may also make smart-phones dearer. Mr White of Bristol York estimates that device makers already have to pay royalties for 200–300 patents for a typical smart-phone. Patent costs are 15–20% of its selling price, or about half of what the hardware components cost. "If 50 people [each] want 2% of a device's value, we have a problem," says Josh Lerner, a professor at Harvard Business School.

Finally, there is a danger that the current intensity of litigation will become normal. Pessimists predict an everlasting patent war, much as the wider information-technology industry seems permanently embroiled in antitrust action. The Wright brothers' legal skirmishes were put to rest only by the outbreak of the first world war. With luck, the smart-phone patent battles will end more quietly.

Clarification: In the infographic, the line between Microsoft and HTC indicates that they have "settled." In fact, Microsoft and HTC entered into a patent licensing agreement—they never litigated over the mobile phones. This clarification was added on October 22nd 2010.

References

Coradi, A., Heinzen, M., & Boutellier, R. (2013). Der SBPC Mix: Schutzrechte für Innovationen. In J. Gausemeier (ed). *Vorausschau und Technologieplanung—9. Symposium für Vorausschau und Technologieplanung.* Heinz Nixdorf Institute, Berlin, Germany.

Ehrat, M. (1997). *Kompetenzorientierte, analysegestützte Technologiestrategieerarbeitung* (Dissertation). HSG, St. Gallen, Switzerland.

Jaffe, A. B., & Lerner, J. (2011). *Innovation and its discontents: How our broken patent system is endangering innovation and progress, and what to do about it.* Princeton University Press, Princeton, New Jersey.

Core Capabilities 16

Making strengths productive.

Peter F. Drucker

Abstract

Core capabilities are the internal strengths that make a company different. They are the product of a long learning process, differentiate the company in the eyes of the customer and cannot be imitated easily. They have to be defined and managed.

In the early 1990s, 3M realized that many households had a problem with rusting soap pad scourers. The company set up a team with members from four divisions, Adhesives, Abrasives, Coatings, and Non-woven, and within one year it was able to come up with a never rusting soap pad scourer made from plastics. A nice side effect, the new pads did not scratch the pans anymore and 3M conquered 30 % of the US market within a year and a half (Coy, Gross, Sansoni, & Kelly, 1994). 3M was able to use the strength of its well-known core capabilities:

- Materials technology
- Product development of small household items
- Marketing: "Produce a little, sell a little", tens of thousands of products

Most companies have their specific core capabilities that make them outstanding in their industry. Amazon.com is strong in short time home delivery through the Internet and is building up a whole range of products around this service. It is the world's largest online retailer. In 1995 Amazon.com started as an online bookstore but soon diversified, selling DVDs, CDs, and MP3 downloads, software, video games, electronics, apparel, furniture, food, toys, and jewellery. Its core capability is "e-tailing", selling online to customers spread all over the world that cannot be reached anymore through traditional expensive distribution chains.

In the 1990s ABB was among the market leaders in building turnkey plants for power production all over the world. Its main core capability was the management of big projects. Mettler Toledo (weighing equipment) and Schindler (elevators) defined their core capabilities explicitly to make sure everybody worked hard to push them further ahead.

In the sense of the resource-based view of the enterprise (Penrose, 1959), a core capability is a "special type of resource", a set of technologies and skills that have been built up over a long period of time and are difficult to imitate. They are customer-oriented. The company focuses on these capabilities and tries to acquire other resources through collaboration with partners that have complementary core capabilities. Essentially, management of core capabilities is the bundling of the resources that build capabilities and ultimately lead to a basis of competitive advantage. These bundles can be "gateways for new opportunities." Strong growth due to mergers & acquisitions often results in relatively autonomous units. In order to use synergies and avoid expensive redundancies, it is important to know exactly what the core capabilities of each unit are.

To create and maintain core technological capabilities, managers need at least two abilities: First, they have to know how to manage the activities that create knowledge and second, they need to know the fundamentals of their core capabilities.

16.1 Managing Core Capabilities

Managing core capabilities needs four steps. The steps cannot be carried out quickly. Time is needed to build up skills and to embed them into the organization.

1. Selection and definition of core capability dimensions: This can be done by carefully searching for products and processes and always asking the questions: Do these skills and technologies provide value to the customers? Can they be protected and are they the gateway to new opportunities as far as the customer is concerned? Do they reduce cost? While selecting core capabilities, the manager should stay focused and not select too many. For small and medium enterprises the suggestions is to choose no more than 10 core capabilities. Usually these core capabilities form a well-balanced net of interwoven resources, policies and actions. They are similar to the strategic advantages according to Rumelt (2011).
2. New knowledge generation: New knowledge can be built by importing knowledge from the outside, experimenting, solving problems and absorbing new methods and new technologies. To institutionalize new knowledge, a learning process with double loop learning is needed. In contrast to single-loop learning, where a specific goal is achieved through different ways, in double-loop learning, goals can be modified or even rejected in the light of experience. People involved in the process discuss not only what they have learnt but also whether it is still worthwhile to learn it or whether new goals should be defined.
3. New knowledge deployment: New knowledge has emerged and it has to be spread over the whole company. This can be done by many different means: meetings, training, projects, role models, constructing stories, etc. It is important to remember that the customer does not buy core capabilities, but products. Thus knowledge has to be transferred into products and processes.
4. Knowledge protection: To keep the competitive advantage, the company has to build up barriers for competitors. If the core capability cannot be protected, it

16.1 Managing Core Capabilities

becomes useless. There are many types of barriers. Intel has factories that cost a fortune. Not everybody has the financial means to build such factories to catch up with such a brand.

IBM, for instance, was not able to protect its PC business: Too much was handed over to Intel and Microsoft. Their operating system was publicly available; their security and distribution systems were standard. Even though in 1983 more than 3,000 IBM software packages were available, they had no sustainable barriers against their competitors. IBMs attempt to come up with its own operating system, OS2 and OS3, failed. MS DOS was already accepted as a standard. In contrast, Nutrasweet built up barriers through patents for its aspartame sweeteners and renewed them regularly. It built up a strong brand with its logo and design. Due to its internal production it was able to cope with copycats.

16.1.1 Internal Strengths and Strategic Values

Core capabilities combine internal strengths and strategic values. Every core capability has its own life. It usually starts with low strategic value and low internal value. In this stage the company can only observe and is not prepared to do anything reasonable at a higher scale. Everything is open. Observation means interpretation for relevance and you can interpret only what you understand. Once the company has acquired some knowledge, pilot projects can be executed and strategic values can be assessed. Generally, the progress is limited by scarce resources, limiting factors, such as people or money. But once pilot projects are finished, the firm has enough knowledge to make larger investments, followed by optimization steps. At the end of its life cycle the core capability has to be divested before it becomes a core rigidity (Leonard-Barton, 1995).

This life cycle, from observation to disinvestment (Fig. 16.1), is a simple tool to present technology-strategy in a one-pager: The company displays what it wants to observe, at which point it goes into feasibility studies, when it moves in with the big money to make investments and finally what it wants to give up.

If the tool is used every year, items move from box to box mostly clockwise, e.g. from investment to optimization. Anti-clockwise may happen as well: A pilot project may show that the company is not yet ready and it moves back to observation.

Experience shows that most companies do not use enough resources for observation and pilot projects. This leads to the well-known investment without adequate preparation with all its risks.

16.1.2 Be Aware of Core Rigidities

The flipside of core capabilities are core rigidities. Core rigidities matter when companies practise over-engineering or become detached from their stakeholders. Then they limit their problem-solving capabilities which leads internally to a sterile

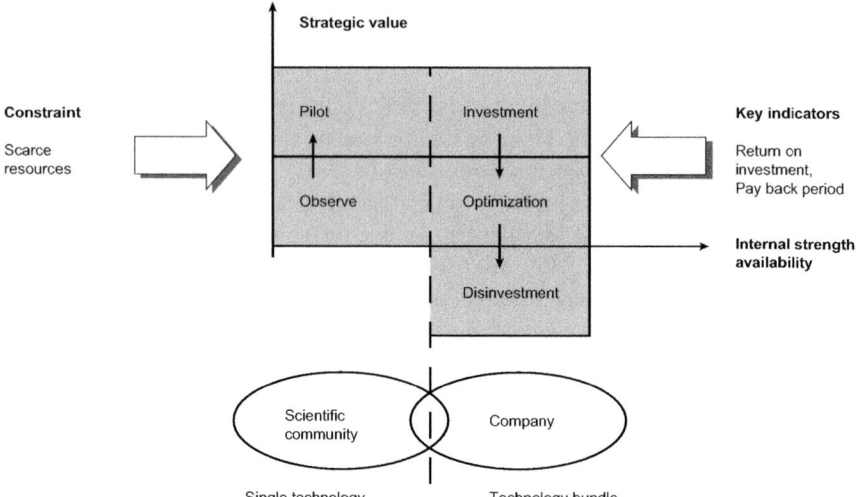

Fig. 16.1 Core capabilities: its specific components have to be managed (Boutellier, Ehrat, & Willemin, 1996)

implementation of solutions and the inability to innovate. Without a sense of creativity and curiosity, experimentation is limited and new knowledge is screened out from the beginning. A vicious circle that is hard to break through again (Fig. 16.2). It needs a revolution according to Kuhn.

There are many reasons why core rigidities evolve. But if they have developed then managers often start fostering them. Many companies discover core rigidities only once they have become obvious when customers leave and market value drops. Then both Board and shareholders realize that a deep change is needed and this can be very costly. This should be avoided from the beginning. Every company needs observation and pilot projects. It has to experiment with new approaches to challenge the existing core capabilities.

One example where core capabilities turned into core rigidities is Kodak. Kodak's strategic intent was to be the "world leader in imaging". This encompasses all kinds of images, delivered by means of technologies such as photography, electrophotography etc. The group identified ten core capabilities, six of which are technical: silver halide imaging materials, non-silver halide imaging materials, precision thin-film coatings and finishing, opto-mechatronics, imaging electronics, and imaging science. Two of these technologies are the ones in which Kodak had invested for decades: silver halide imaging started with the origins of the company in 1881 and imaging science research started in 1913 with the Research Laboratories of Kodak. Imaging electronics is a much newer field and like Polaroid, Fuji and others, Kodak was struggling with developing the required knowledge. The strategic intent of Kodak, however, provided a clear and rigid guideline without flexibility and led to a disaster in the end. Kodak missed the opportunity to change horses early enough. Kodak filed for bankruptcy protection in January 2012.

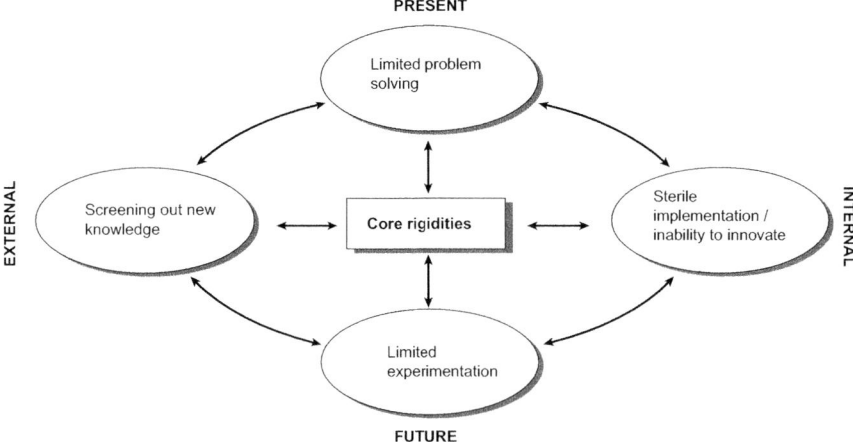

Fig. 16.2 Knowledge-inhibiting activities: core rigidities

16.1.3 Skills and Problem-Solving

Problem-solving is supported and constrained at the same time through prior experiences, practices and most important success. In fact, people easily presume to solve problems innovatively once they have established their habits. The phenomenon underlying this seems to be the brain's natural tendency to store, process, and retrieve information in related blocks. These mind-sets are highly useful in routine activities as they lead to efficient problem-solving. However, the problem can be that the limited range of problem-solving responses can contribute, as seen before, to core rigidities. A person highly skilled in specific methods frames problems in such a way that they can be solved with those familiar and preferred methods. "Give a little boy a hammer and everything looks like a nail!" Managers should select some employees particularly because their ideas, personalities, values and skills conflict with the standard rules and values of the company.

Gerald Hirshberg, former director of Nissan Design International, coined the phrase "creative abrasion" to emphasize that energy created by conflicts can be channelled into creativity rather than destruction (Fig. 16.3). Such managers build an atmosphere that encourages people to respect each other's viewpoint without necessarily agreeing with it. But it is management's task to ensure that friction becomes creative.

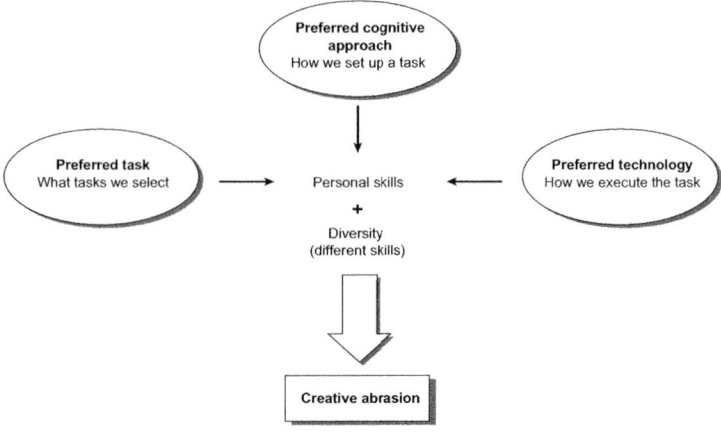

(Hirshberg, former director of Nissan Design International)

Fig. 16.3 Personal skills = routine combined with diversity triggers creativity

16.2 Learning

Individual and organizational learning is at the base of every adaptive organization and is needed to create core capabilities and to avoid core rigidities. In this sense, double loop learning is essential, because not only problem solutions are questioned, but assumptions and goals as well (Fig. 16.4).

Therefore single-loop learning refers to incremental improvements and problem-solving within an existing set of rules and norms, while double-loop learning involves changes in the fundamental rules and values underlying action and behaviour and has consequences for the whole organization.

As in routine processes, learning in companies occurs through repeated cycles of plan, do, check, act (PDCA). Deming's PDCA-cycle can be also transferred to a technology life cycle that may lead to a creation of technical core competencies.

Organizations need both single and double-loop learning to cope both with the challenges of the everyday order-make delivery-process and future opportunities. In continuous cycles of learning targets and solutions the "What" and the "How" are developed in parallel, not consecutively. The difference between single and double loop learning is similar to the difference between the rational approach to strategy and the emergent strategies. Rational means: Make an assessment of the situation, choose a strategy and make it happen. Emergent means: Keep options open, choose a structure that allows a strategy to emerge over time and do not separate realization from strategy development. The latter becomes more important in cases of high uncertainty.

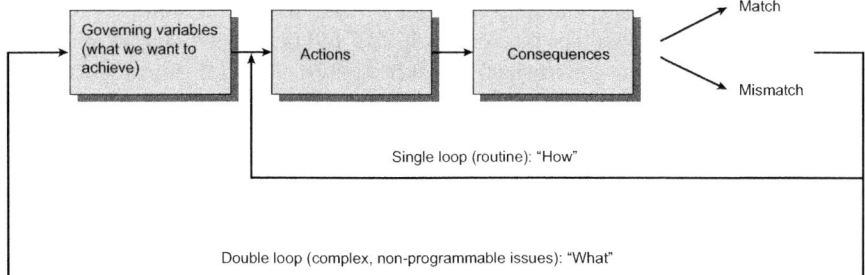

Fig. 16.4 Organizations need single loop and double loop learning: both need matches/mismatches (Argyris, 1999)

The Economist: Technological change: The last Kodak moment?

Kodak is at death's door; Fujifilm, its old rival, is thriving. Why?

© The Economist Newspaper Limited, London (Jan 14th 2012)

LENIN is said to have sneered that a capitalist will sell you the rope to hang him. The quote may be spurious, but it contains a grain of truth. Capitalists quite often invent the technology that destroys their own business.

Eastman Kodak is a picture-perfect example. It built one of the first digital cameras in 1975. That technology, followed by the development of smartphones that double as cameras, has battered Kodak's old film- and camera-making business almost to death.

Strange to recall, Kodak was the Google of its day. Founded in 1880, it was known for its pioneering technology and innovative marketing. "You press the button, we do the rest," was its slogan in 1888.

By 1976 Kodak accounted for 90% of film and 85% of camera sales in America. Until the 1990s it was regularly rated one of the world's five most valuable brands.

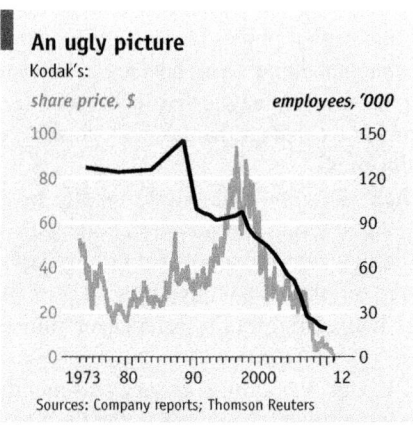

Then came digital photography to replace film, and smartphones to replace cameras. Kodak's revenues peaked at nearly $16 billion in 1996 and its profits at $2.5 billion in 1999. The consensus forecast by analysts is that its revenues in 2011 were $6.2 billion. It recently reported a third-quarter loss of $222m, the ninth quarterly loss in three years. In 1988, Kodak employed over 145,000 workers worldwide; at the last count, barely one-tenth as many. Its share price has fallen by nearly 90% in the past year (see chart).

For weeks, rumours have swirled around Rochester, the company town that Kodak still dominates, that unless the firm quickly sells its portfolio of intellectual property, it will go bust. Two announcements on January 10th—that it is restructuring into two business units and suing Apple and HTC over various alleged patent infringements—gave hope to optimists. But the restructuring could be in preparation for Chapter 11 bankruptcy.

While Kodak suffers, its long-time rival Fujifilm is doing rather well. The two firms have much in common. Both enjoyed lucrative near-monopolies of their home markets: Kodak selling film in America, Fujifilm in Japan. A good deal of the trade friction during the 1990s between America and Japan sprang from Kodak's desire to keep cheap Japanese film off its patch.

Both firms saw their traditional business rendered obsolete. But whereas Kodak has so far failed to adapt adequately, Fujifilm has transformed itself into a solidly profitable business, with a market capitalisation, even after a rough year, of some $12.6 billion to Kodak's $220m. Why did these two firms fare so differently?

Both saw change coming. Larry Matteson, a former Kodak executive who now teaches at the University of Rochester's Simon School of Business, recalls writing a report in 1979 detailing, fairly accurately, how different parts of the market would switch from film to digital, starting with government reconnaissance, then professional photography and finally the mass market, all by 2010. He was only a few years out.

Fujifilm, too, saw omens of digital doom as early as the 1980s. It developed a three-pronged strategy: to squeeze as much money out of the film business as possible, to prepare for the switch to digital and to develop new business lines.

Both firms realised that digital photography itself would not be very profitable. "Wise businesspeople concluded that it was best not to hurry to switch from making 70 cents on the dollar on film to maybe five cents at most in digital," says Mr Matteson. But both firms had to adapt; Kodak was slower.

A culture of complacency

Its culture did not help. Despite its strengths—hefty investment in research, a rigorous approach to manufacturing and good relations with its local community—Kodak had become a complacent monopolist. Fujifilm exposed this weakness by bagging the sponsorship of the 1984 Olympics in Los Angeles while Kodak dithered. The publicity helped Fujifilm's far cheaper film invade Kodak's home market.

Another reason why Kodak was slow to change was that its executives "suffered from a mentality of perfect products, rather than the high-tech mindset of make it, launch it, fix it," says Rosabeth Moss Kanter of Harvard Business School, who has advised the firm. Working in a one-company town did not help, either. Kodak's bosses in Rochester seldom heard much criticism of the firm, she says. Even when Kodak decided to diversify, it took years to make its first acquisition. It created a widely admired venture-capital arm, but never made big enough bets to create breakthroughs, says Ms Kanter.

Bad luck played a role, too. Kodak thought that the thousands of chemicals its researchers had created for use in film might instead be turned into drugs. But its pharmaceutical operations fizzled, and were sold in the 1990s.

Fujifilm diversified more successfully. Film is a bit like skin: both contain collagen. Just as photos fade because of oxidation, cosmetics firms would like you to think that skin is preserved with anti-oxidants. In Fujifilm's library of 200,000 chemical compounds, some 4,000 are related to anti-oxidants. So the company launched a line of cosmetics, called Astalift, which is sold in Asia and is being launched in Europe this year.

Fujifilm also sought new outlets for its expertise in film: for example, making optical films for LCD flat-panel screens. It has invested $4 billion in the business since 2000. And this has paid off. In one sort of film, to expand the LCD viewing angle, Fujifilm enjoys a 100% market share.

George Fisher, who served as Kodak's boss from 1993 until 1999, decided that its expertise lay not in chemicals but in imaging. He cranked out digital cameras and offered customers the ability to post and share pictures online.

A brilliant boss might have turned this idea into something like Facebook, but Mr Fisher was not that boss. He failed to outsource much production, which might have made Kodak more nimble and creative. He struggled, too, to adapt Kodak's "razor blade" business model. Kodak sold cheap cameras and relied on customers buying lots of expensive film. (Just as Gillette makes money on the blades, not the razors.) That model obviously does not work with digital cameras. Still, Kodak did eventually build a hefty business out of digital cameras—but it lasted only a few years before camera phones scuppered it.

Kodak also failed to read emerging markets correctly. It hoped that the new Chinese middle class would buy lots of film. They did for a short while, but then decided that digital cameras were cooler. Many leap-frogged from no camera straight to a digital one.

Kodak's leadership has been inconsistent. Its strategy changed with each of several new chief executives. The latest, Antonio Perez, who took charge in 2005, has focused on turning the firm into a powerhouse of digital printing (something he learnt about at his old firm, Hewlett-Packard, and which Kodak still insists will save it). He has also tried to make money from the firm's huge portfolio of intellectual property—hence the lawsuit against Apple.

At Fujifilm, too, technological change sparked an internal power struggle. At first the men in the consumer-film business, who refused to see the looming crisis, prevailed. But the eventual winner was Shigetaka Komori, who chided them as "lazy" and "irresponsible" for not preparing better for the digital onslaught. Named boss incrementally between 2000 and 2003, he quickly set about overhauling the firm.

Mount Fujifilm

He has spent around $9 billion on 40 companies since 2000. He slashed costs and jobs. In one 18-month stretch, he booked more than ¥250 billion ($3.3 billion) in restructuring costs for depreciation and to shed superfluous distributors, development labs, managers and researchers. "It was a painful experience," says Mr Komori. "But to see the situation as it was, nobody could survive. So we had to reconstruct the business model."

This sort of pre-emptive action, even softened with generous payouts, is hardly typical of corporate Japan. Few Japanese managers are prepared to act fast, make big cuts and go on a big acquisition spree, observes Kenichi Ohmae, the father of Japanese management consulting.

For Mr Komori, it meant unwinding the work of his predecessor, who had handpicked him for the job—a big taboo in Japan. Still, Mr Ohmae reckons that Japan Inc's long-term culture, which involves little shareholder pressure for short-term performance and tolerates huge cash holdings, made it easier for Fujifilm to pursue Mr Komori's vision. American shareholders might not have been so patient. Surprisingly, Kodak acted like a stereotypical change-resistant Japanese firm, while Fujifilm acted like a flexible American one.

Mr Komori says he feels "regret and emotion" about the plight of his "respected competitor". Yet he hints that Kodak was complacent, even when its troubles were obvious. The firm was so confident about its marketing and brand that it tried to take the easy way out, says Mr Komori.

In the 2000s it tried to buy ready-made businesses, instead of taking the time and expense to develop technologies in-house. And it failed to diversify enough, says Mr Komori: "Kodak aimed to be a digital company, but that is a small business and not enough to support a big company."

Perhaps the challenge was simply too great. "It is a very hard problem. I've not seen any other firm that had such a massive gulf to get across," says Clay Christensen, author of "The Innovator's Dilemma", an influential business book. "It was such a fundamentally different technology that came in, so there was no way to use the old technology to meet the challenge."

Kodak's blunder was not like the time when Digital Equipment Corporation, an American computer-maker, failed to spot the significance of personal computers because its managers were dozing in their comfy chairs. It was more like "seeing a tsunami coming and there's nothing you can do about it," says Mr Christensen.

Dominant firms in other industries have been killed by smaller shocks, he points out. Of the 316 department-store chains of a few decades ago, only Dayton Hudson has adapted well to the modern world, and only because it started an entirely new business, Target. And that is what creative destruction can do to a business that has changed only gradually—the shops of today would not look alien to time-travellers from 50 years ago, even if their supply chains have changed beyond recognition.

Could Kodak have avoided its current misfortunes? Some say it could have become the equivalent of "Intel Inside" for the smartphone camera—a brand that consumers trust. But Canon and Sony were better placed to achieve that, given their superior intellectual property, and neither has succeeded in doing so.

Unlike people, companies can in theory live for ever. But most die young, because the corporate world, unlike society at large, is a fight to the death. Fujifilm has mastered new tactics and survived. Film went from 60% of its profits in 2000 to basically nothing, yet it found new sources of revenue. Kodak, along with many a great company before it, appears simply to have run its course. After 132 years it is poised, like an old photo, to fade away.

References

Argyris, C. (1999). *On organizational learning.* Oxford: Blackwell.
Boutellier, R., Ehrat, M., & Willemin, J.-F. (1996). Konzentration der Kräfte: Voraussetzung für Innovation und Wachstum: Technologiestrategien an Kernkompetenzen orientiert erarbeiten. *VDI-Z, 6*(138), 34–38.
Coy, P., Gross, N., Sansoni, S., & Kelly, K. (1994). What is the world in the lab? Collaborate. *Business Week, 27,* 44–45.
Leonard-Barton, D. (1995). *Wellsprings of knowledge: Building and sustaining the sources of innovation.* Boston, MA: Harvard Business Press.
Penrose, E. (1959). *The theory of the growth of the firm.* New York, NY: Wiley.
Rumelt, R. (2011). *Good strategy bad strategy: The difference and why it matters.* New York, NY: Random House Digital.

Technical Risk Management 17

> *Risk comes from not knowing what you are doing.*
> Warren Buffet

Abstract
Technical failures have become routine in every-day work life. Shorter development times, an increasing diversity in applications and a much higher complexity are leading to high failure rates for cars and many other mass products. Reliability has to be designed into the product. To prioritize scarce resources, engineers apply Mean Time Between Failure as the most common quantitative tool.

17.1 Construction Failures

17.1.1 The Hyatt Regency, USA

The Hyatt Regency Hotel in Kansas City, Missouri, opened in July of 1980 and one year later the walkways on the second and the fourth floor collapsed during a local radio station's dance competition, killing 114 people and injuring 200. It was the worst structural failure in the history of the United States until the Collapse of the World Trade Centre in September 2001.

The lobby was one of the most attractive features of the hotel with its atrium spanned by steel, glass and concrete walkways suspended from the ceiling on the second, third and fourth floors (Kaminetzky, 1991). Each of the three walkways was 37 m long and weighed 29,000 kg. The fourth level walkway was aligned directly above the second level walkway. A major cause of the fatalities was the crash of this concrete fourth floor walkway onto the crowded second floor walkway. On that day, approximately 40 people were standing on the second level walkway and 20 on the fourth level walkway. They all fell onto the crowd in the lobby.

What happened? Three days after the disaster, Wayne Lischka, a structural engineer hired by The Kansas City Star newspaper, found that the actual construction of the walkways had one significant difference compared with the original design. His publications on the event even earned him a Pulitzer Prize for local news reporting in 1982. During construction an engineer had changed a

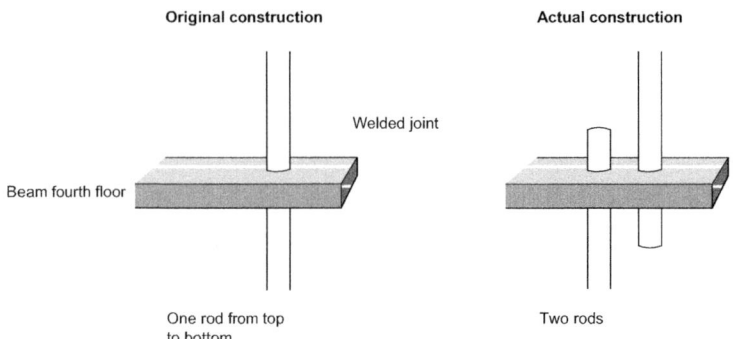

Fig. 17.1 Hyatt Regency, planned versus built rod connection

construction "detail": Instead of hanging the walkways onto 6 vertical tension rods going from the ceiling down to the lowest floor walkway, he used 12 tension rods measuring half the required length. The upper rod carried the fourth floor and the second rod all the lower floor walkways. The two rods were connected through the beam under the fourth floor walkway. However, he did not change these connections accordingly. The beam on the fourth floor had to carry far more load than originally specified (Fig. 17.1).

This design change was fatal. In the original design, the beams underneath the fourth floor walkway had to support only the weight of the fourth floor walkway. In the revised design, however, the fourth floor beams were required to support both the fourth floor walkway and the second floor walkway. With the load on the fourth floor beams doubled, the new construction carried only 30 % of the weight required.

The serious mistakes of the revised design were made even worse by the fact that both designs placed the bolts directly through a welded joint connecting two channels, the weakest structural point in the box beams. Photographs of the wreckage show excessive deformations of the cross-section. During the collapse, the box beams split along the weld and the nut supporting them slipped through the resulting gap between the connections which had been welded together.

The investigators concluded that the basic problem was a lack of proper communication between the engineers. However, during the investigation, all of the involved engineers testified that they had notified the responsible engineer on all construction changes and that he had given a confirmation. The engineers who had approved the final drawings were convicted of unprofessional conduct in the practice of engineering; they all lost their engineering licenses. The system of construction planning was not changed; only some people lost their jobs. This is the standard reaction to disasters: Remove the responsible people and keep the old procedures. Construction failures are not uncommon, unfortunately.

17.1.2 Indoor Swimming Pool in Uster, Switzerland

Another disaster happened in Uster near Zurich, Switzerland, that led to research which changed our understanding of the phenomenon of stress corrosion cracking – a phenomenon in stainless steel. In 1972 Uster inaugurated its new indoor swimming pool. An 80 mm thick concrete ceiling was suspended underneath the roof of the building. It was fixed with 207 chrome nickel steel bars, a well-known stainless steel. In 1981 an additional wood ceiling was added to improve the acoustics. In 1985 at 8 o'clock in the evening a loud bang came from the ceiling and 15 min later the ceiling fell down. The crash drowned 12 people. A subsequent inspection showed that material and construction met all of the specifications, but 30 rods were completely corroded and 25 rods were 75 % corroded. The safety factor went down from the required 1.8 to the sub-optimal 1.3.

Stress corrosion cracking occurs when we have:
- Tensile stress
- High temperature
- A corrosive environment
- A susceptible material

Until the accident in Uster, all engineering evidence suggested that stress corrosion cracking in standard stainless steels did not occur below 50 °C. Although pools have become warmer over the years, they do not reach the "traditional" 50 °C mark. When corrosion was diagnosed in Uster, this came as somewhat of a shock to the corrosion science community. Construction experts did not know that stainless steel would corrode in the high chlorine concentration environment of a swimming pool. Stainless steel breaks, but ordinary steel enters an elastic phase before breaking and protects itself with rust. With ordinary steel the accident would probably not have happened! Ten years after the incident in Uster, another swimming pool was built in Switzerland with the same defects. After the design was finished, fortunately somebody heard about Uster and the design of the ceiling was changed! The accident in Uster shows that accidents with new materials may happen years after construction, when legal limitation periods have run out.

17.2 Nuclear Disasters

Until the disastrous earthquake on March 14th 2011 in Japan that killed more than 20,000 people and the following nuclear disaster in Fukushima, we had had two major accidents involving nuclear energy: Three Mile Island and Chernobyl. The professionals in the Manhattan WW II bomb project were very much traumatized by Hiroshima and Nagasaki and took all precautions in order to prevent an accident in civilian applications. During the development of nuclear technology, safety was ensured on the run until President Carter stopped the whole program. Nuclear power technology has a peculiar history. After the war, everybody was looking for a new power source for submarines. Traditional submarines were being driven by diesel engines and had to surface every few hours for oxygen. With nuclear power they could stay underwater for weeks. Back then the military was only interested in technology that could be miniaturized. Since then today's boiling water reactor has evolved. It is the only reactor that was built in large numbers with a civilian use in mind. Thus the technology we use today was not chosen because it is safe, but because it was the only technology that was field-tested for use in submarines. There are safer designs but nobody has had the courage to spend the billions needed to test them.

At the beginning, civilian nuclear technology was confined to remote places and only used on a low scale. But transporting electricity over long distances is expensive, thus engineers devised the containment concept: Radioactivity and heat were contained in a strong steel case. Our confidence in nuclear energy started to grow. However, when the first 1,000 MW station was built, people started to worry that a total core meltdown would occur which, in our fantasy, could burn a hole all the way to the other side of the earth. Thus the "China Syndrome" was born in a movie with Jane Fonda and Jack Lemmon that challenged the safety of nuclear power. It started in US cinemas just 12 days before Three Mile Island happened.

In order to avoid accidents, engineers added backup systems which would be able to cool the core of the reactor. Public hearings in the US showed concerns of a breakdown in the backup system. Everybody finally accepted that there would always be a residual risk. Engineers tried to show that this risk was very small. Professor Rasmussen from MIT made an estimate and used a probabilistic approach to risk assessment that is still in use today. Rasmussen released his report in 1975. He estimated the probability of components of the system failing and calculated the probability of a core damage accident. This number turned out to be extremely small: Only once every 20,000 years of operating time! Four years later, on March 28th, 1979 the accident at Three Mile Island happened after only 500 years of operating time. Even more backups were added and the US nuclear program was killed off due to the costs of all the safety measures.

Safety management went through three phases:
- In the 1950s reliability theory and project management were used to overcome growing complexity.
- In the 1970s the big question was how safe is safe enough, the probabilistic approach was born.
- In the 1980s after Three Mile Island, technical risk management was questioned more and more. A public discussion started that has continued since then.

In the twenty-first century, new technical safety measures have been introduced and more and more passive backup systems have been installed. Old systems are ramped up to the newest safety levels: Beznau II, a Swiss nuclear power station, cost CHF 700 million and in 2010 its technical improvements about CHF 1,400 million. These additional investments were mostly driven by safety, not efficiency.

17.3 Engineering Approach to Risk

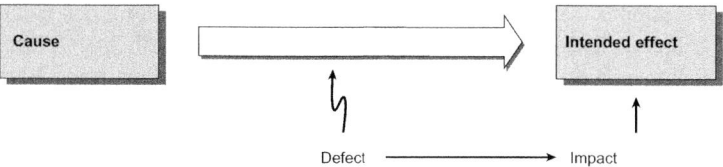

Fig. 17.2 Engineers have a very specific understanding of technical risks: they want to calculate

17.3 Engineering Approach to Risk

Engineers have a very specific understanding of technical risks (Fig. 17.2): A system designed to achieve a certain goal can have a defect with a specific impact and a specific likelihood of occurrence.

As engineers want to make comparisons and be able to set priorities, they express this idea in terms of numbers: One has to estimate the probability of defect occurrence and quantify the impact with an estimate of its resulting costs. In order to calculate the risk of a certain event, its probability and cost are multiplied by each other:

$$\begin{aligned} Risk &= statistical\ expectation \\ &= probability\ that\ the\ defect\ occurs * impact\ (cost) \\ &= a\ cost \end{aligned}$$

In statistical terms this is an expectation and it is measured in cost (USD). In order to make the decision of whether to introduce pedestrian crossings or traffic lights, we have to calculate the probability for both approaches. These probabilities then need to be multiplied by the cost of the casualties.

The same procedure can be used to calculate the risk of a fatal accident on a car journey. We know that Switzerland has some 400 deadly casualties per year in car crashes and some 1,200,000,000 km are driven in the country.

$$km\ driven * \frac{400}{1,200,000,000} * Value\ of\ life = \frac{Expected\ loss}{Drive}$$

As there will be fatal and non-fatal accidents, a number has to be put on the value of life (Lewis, 1990): Most people think this is cynical! However, since society has limited resources and since traffic is an accepted technology, engineers and administrators have to make such decisions to compare different technologies and set priorities. Once an accident has happened, the value of life is infinite. Nobody asked how much it would cost to save the life of a fireman trapped in the rubble of the underground car park in Gretzenbach, Switzerland in 2004 after it collapsed and killed seven firemen. The question was simply how many resources could be used in parallel. Thus it is important to differentiate between the calculations made before the accident and the actual reaction after the accident. But we understand from this example why so many people have problems accepting risk calculations as soon as human lives are involved. It makes a difference whether you calculate the risk of losing some money on a financial market or whether some people get severely hurt.

17.4 Technical Risk Theory

Technical risk management relies heavily on mathematics and thus simplification. The simplifications are needed to overcome complexity and problem measurement. Usually the simplifications are chosen so that the resulting numbers show the worst case.

17.4.1 The Hazard or Failure Rate

Most calculations start with the hazard rate, the probability that a failure occurs at time t. It is assumed to be constant during the useful lifetime of a system. This is a first simplification. Car crashes are more frequent in a fog than on a clear day.

The probability of failure has three phases (Fig. 17.3):

1. The first phase is a decreasing failure rate known as early failures or "teething troubles". At the beginning after the introduction of a new product, unexpected and sporadic failures can occur. New systems cannot be tested long enough; they would be on the market too late. Thus even well-known systems like iPhones have some problems at the beginning.
2. The second part is a constant failure rate known as random failures in the granted utilization time. Through periodic maintenance it can be kept more or less constant. You have to change the oil in your car every 10,000 km.
3. The last phase is characterized by an increasing failure rate known as wear-out failures due to wear and tear when predictability is lost again. When your car gets very old you cannot predict anymore when you will next have to go to your garage.

The cross-sectional shape of a bathtub gave the curve its name and is widely used in reliability engineering. Many consumer products strongly reflect the bathtub curve such as computer processors, cars and flat screens. In the pharmaceutical sector, laws require extended trials, thereby driving the cost of new drugs into the billions and incurring fatalities as well as a result of the introduction of the drug being delayed. Drugs do not wear out, but more and more side-effects emerge which may limit the original market segment.

The goal is to keep the failure rate constant for as long as possible. This goal can be reached with improvements such as endurance tests in gas turbines, process controls in light bulbs or a rugged design in climbing ropes.

While the bathtub curve is useful, not every product or system follows a bathtub curve; for example, if units have a decreased use during or before the onset of the wear-out period, they will show fewer failures per unit calendar time (not per unit

17.4 Technical Risk Theory

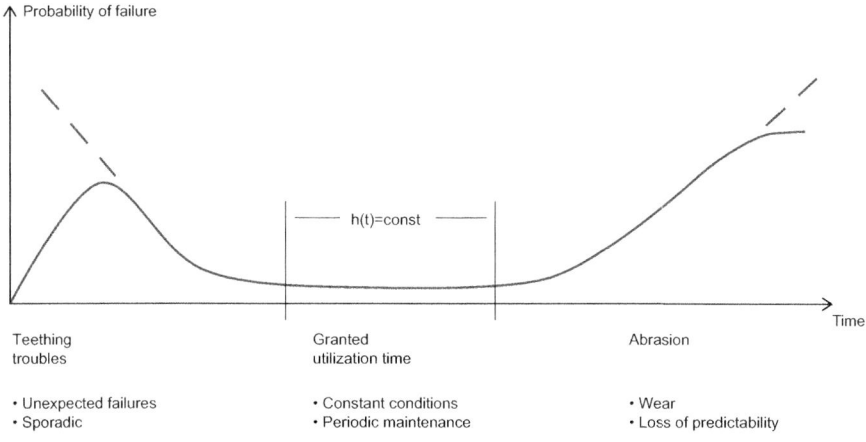

Fig. 17.3 The U – principle of reliability ("Bathtub curve")

use time) than the average bathtub curve shows. Thus the bathtub curve also has its critics. Some investigations in aerospace and other complex products have revealed that many failures do not comply with it. It is argued that the bathtub curve is an old concept and should not be used as a stand-alone guide to reliability.

17.4.2 Reliability

If we are interested as to how long we can wait on average until we have a failure, we do not use the bathtub curve, we use the probability of there being no failure until time t. R (100 h) = 0.5 which means that 50 % of all products have no failure in the first 100 h.

When the probability of failure is constant, simple mathematics shows that reliability drops off exponentially and quickly, with time.

In a further approximation, the reliability of the system is the probability that none of its components will fail, that everyone will work. And this is equal to the product of all reliabilities, another simple rule of probability theory. Thus and this is the clue for all calculations, when we know the reliability of the components we can calculate the reliability of the system.

This is a second simplification because we assume that there are no connections between the different components: All failure types are assumed to be independent and lead to a failure of the system. But if you ride a bicycle with a flat tyre for a few kilometers, your spokes will suffer, their reliability will be affected and the total system will behave differently. As soon as cross-influences are taken into account, the results cannot be measured anymore and we are forced to use theoretical models with no experimental backup! There are too many potential combinations. A problem we meet whenever we calculate tolerances for technical systems.

17.4.3 Mean Time Between Failures (MTBF)

To make things more practical, engineers use the inverse of the probability that we have a failure at time t: They use the Mean Time Between Failures, the MTBF, at the component and system level. MTBF is the average time between two failures. Like reliability, the MTBF of a system can be calculated from the MTBFs of the components.

$$MTBF\ (System) = 1/\left(\frac{1}{MTBF1} + \frac{1}{MTBF2} + \ldots + \frac{1}{MTBFn}\right)$$

In other words, the MTBF is the predicted elapsed time between the failures of a system during operation. An MTBF of 1,000 h, typical for our old electrical lamps, means that, on average, lamps survive the first 1,000 h. If a hotel has 50 lamps burning in its dining room for 20 h per day on average, then hotel service has to change one lamp every day. You can easily check whether the hotel has reliable facility management or not. If they started using modern LED lamps, they would only have to replace one lamp per month! This is the major reason why it is cheaper to have LEDs in street lamps.

The definition of MTBF depends on the definition of a system failure. For complex, repairable systems, a failure is defined as a condition or state which renders the system out of order and necessitates a shutdown of operation. Say the orange lamp in your car lights up. This is a condition which does not render the system out of order and does not warrant a repair and is not considered to be a failure, even if a component did break or was damaged. Units or components which need to undergo routine scheduled maintenance or inventory control are not considered to have failed either. Your car comes under this classification: An orange warning lamp tells you that something is wrong but that a repair is not immediately necessary and that you can continue driving. A red lamp tells you to go to the next garage immediately.

This is the theory Professor Rasmussen applied in his famous report about the safety of nuclear power stations. And it is the theory still being used to calculate the reliability of space objects such as shuttles and satellites. Sometimes it works. It is a valuable tool to set priorities for the improvement of a system. System suppliers in the car industry or space industry need to provide these numbers with each new product and each modification.

17.5 Limitations of the Technical Risk Approach

There are three technical problems with this concept:
1. It assumes the worst case, which means that each component failure implies a system failure.
2. The measurement of the MTBF of the components takes time. You should test many components until they fail. This is no problem for traditional lamps. But

who waits 20,000 h, 2½ years, for a new LED Lamp before you enter the market? As there is not enough time, estimates are made – simplifications, because they rely on small sample sizes and short observation times.
3. It is not the worst case because it makes the assumption that the components are independent of each other. Most catastrophes are triggered by a series of failures or "half failures". A system may fail without any single module failing completely. During hot summer days, air-conditioning systems in trains break down regularly. They are not repaired: The system simply runs on overload and recovers as soon as the weather gets cooler.

Nevertheless, the concept has been applied many times and is valuable during the design phase of a product. It gives the engineer a simple tool to calculate the probabilities of failure and thus gives priorities for improvement: Reliability has to be designed. At the beginning of a project, tolerances are usually very tight. The more experienced designers and production managers become, the less tight tolerances cost. When component tolerances can be reduced, reliability goes up. Engineers have a tolerance budget and the allocation of the tolerances has a substantial cost impact. We see this with the falling prices of CPUs when they reach the market. At the beginning they have a high waste rate that goes down with experience and continuous improvement.

More important objections to the MTBF concept are:
- It does not consider side-effects. The example of asbestos shows that the worst risks are sometimes unexpected side-effects, seen decades after use!
- It says nothing about the perception and acceptance of risk. People do not like the worst-case scenario of a nuclear power plant with presumably a few hundred dead, even if the probability is something like 0.00000... per year. Low probability does not mean impossible!
- People would like to have risks distributed according to some principle perceived as just. MTBF says nothing about this distribution.
- MTBF works only for "inanimate objects", for machines. As soon as humans are involved, we have non-linearities, people react as soon as they perceive risks. Systems become adaptive and thus chaotic: For some hours hundreds of control lamps went red in the control room at Three Mile Island. Nobody was able to give a reasonable interpretation and so the system ran out of control.

We have to accept that MTBF is just part of an overall risk management package. It is powerful and helps us to set priorities and reduce risks.

The Economist: Daily chart: Danger of death!

© The Economist Newspaper Limited, London (Feb 14th 2013)
   How you are unlikely to die
   ON FEBRUARY 15th DA14, an asteroid 45 m across, will sail past the Earth at 7.8km a second (4.9 miles a second). At just 27,700km away, it is well within the range of communication satellites. It will be the closest encounter on record with an asteroid this big. In 1908 an asteroid estimated to be around 100 m in diameter

destroyed 2,000km^2 of forest in Siberia. Thankfully, such events are rare. NASA has identified 9,600 "near-Earth objects" since 1995, but just 861 with a diameter of 1km or more. The greatest threat to Earth currently is the 130-metre wide 2009 FD; but it has just a 1-in-526 chance of hitting the planet, and not until March 29th 2185. More prosaic things are far more dangerous. According to data from America's National Safety Council, 27 people died in 2008 in America from contact with dogs (a one in 11m chance of death). The chart below compares the odds of dying in any given year from choking, cycling, being struck by lightning or stung by a bee.

The Economist: Daily chart: Danger of death!

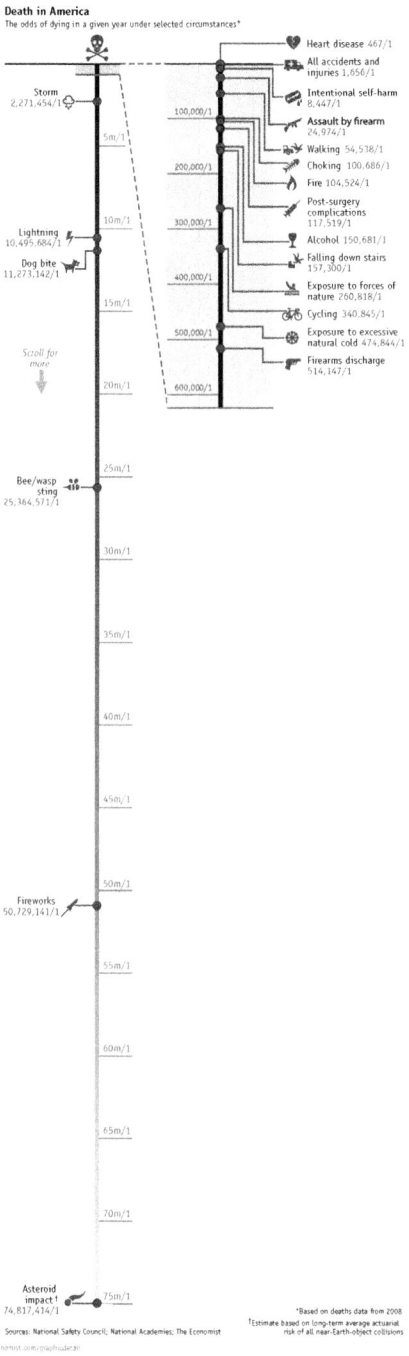

Correction: We originally identified asteroid AG5 as posing the greatest threat to the planet. But as keen asteroid-watchers know, that rock was given the all-clear by NASA in December 2012. The text was corrected on February 20th.

References

Kaminetzky, D. (1991). *Design and construction failures: Lessons from forensic investigations.* New York, NY: McGraw-Hill.

Lewis, H. (1990). *Technological risk.* New York, NY: Norton.

Human Aspects of Technical Risk Management

18

> *Bounded rationality makes risk management possible; we are able to convince, to lead a risk dialogue, to overcome complex situations.*
> Herbert Simon (1916–2001), American economist

Abstract

Technology has never been accepted as broadly as it is today. There are three exceptions, however: Nuclear, bio and gene technology. Whether a technology is accepted or not depends less on rational arguments and more on whether we trust the people who run and supervise the technology. As new technology has increased manifold, there is no time to test everything in detail and science does not give us the answers about threshold levels and safety limits. Risk acceptance cannot be calculated. Instead, it has to be discussed in a democratic dispute.

In 1988, 26 years after the introduction of the first breast implants filled with silicone gel, the U.S. Food and Drug Administration (FDA) investigated breast implant failures and their complications. It required manufacturers to present the necessary documentation confirming the safety of their breast implants. In 1992 the FDA suspended the use of silicone for most non-medical applications. Concerns had been around for many years: More than 10,000 women in the United States filed lawsuits against the producers of silicone implants. Dow Corning Corporation, the biggest silicone product and breast implant manufacturer, agreed to pay USD 4.25 billion for an out of court settlement, even though there was very little incriminating evidence against the company, but it faced 19,000 breast implant sickness lawsuits in the US alone. A further medical study conducted in 1994 showed there was no increased risk of developing complications from silicone implants. However, the litigation went on anyway and in 1995 Dow Corning filed for bankruptcy, under the pressure of 400,000 women asking for compensation. In 2004 the legal case was closed with Dow paying more than USD 4 billion in settlements.

In 2007 there were 350,000 breast augmentations in the US. This case shows the power of class actions in the U.S. and what happens to big companies once public opinion turns against them and customers feel betrayed.

18.1 Disputed Technologies

Today we live in an era of technology euphoria. There are three big technologies that are disputed at the moment:
1. Gene technology: Outside the US, Brazil and China there is a general rejection of genetically modified food. We have a dilemma: Many scientists see the length of field tests as the only difference between a natural plant and a genetically modified plant, and argue that the wild type has the longest test and should therefore be more robust than any of its artificial mutations. If anything goes wrong, the wild type should survive.
2. Biotechnology: Many people are reluctant to accept the patenting of animals and are afraid of potential damage. Even biotechnical processes that yield the same results as proven traditional ones are being questioned. However, beer has been produced by biotechnical methods for a few thousand years. Biotech is such a large field that we have to differentiate.
3. Nuclear technology: Waste disposal and the general risks of this technology have prevented new power stations from being built in many countries. The accident at Fukushima in Japan has slowed down the revival of nuclear power for many years to come. Some countries have decided to move out of this technology for good.

Today's public opinion about nuclear technology is polarized and driven by deep emotions. After Three Mile Island, Chernobyl and Fukushima people do not believe that the probability of the maximum credible accident is close to zero. The proliferation of adverse technology has increased the uncertainty surrounding nuclear energy and the issue of waste disposal is still a hot political issue in most free countries. "Kastor" transports in Germany will continue to cause political unrest.

18.2 Risk Perception

New technologies have initiated big economies of scale. The associated risks have been reduced substantially due to the experience gained over decades. Despite some potentially dangerous technologies, life prospers. However, diseconomies of scale can also occur: Big technologies have become very complex. No single individual fully understands how a Boeing 737 works or what the function of every single component of a big chemical plant is: This is one of the sources of uncertainty. Whenever anxiety driven by a technology coincides with wealth, emotions overcome rational judgement.

Society should not be surprised by these apparent inconsistencies. Many catastrophes have arisen in a climate of increased confidence that was supported by common sense and experience (Fig. 18.1).

New innovations, such as suspension bridges, work perfectly well on a small scale. A climate of confidence arises which verifies the hypothesis that a suspension bridge is a valuable type of bridge. This leads to more and more suspension bridges

18.2 Risk Perception

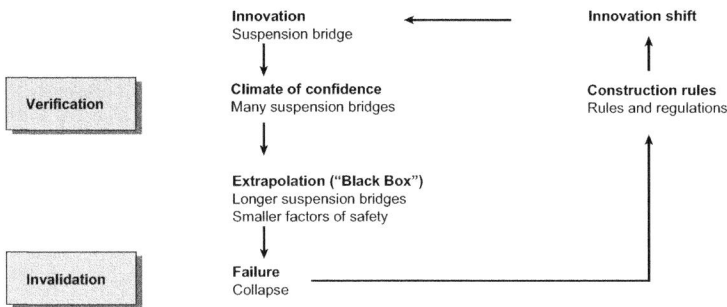

Fig. 18.1 Catastrophes arise in a climate of "increased confidence"

being built and as long as they all perform as predicted, confidence rises. With each new bridge, engineers experiment with new features and tend to test the limits by playing with size and safety factors. This expansion usually lasts until the hypothesis is challenged and invalidated and a suspension bridge collapses. This, in turn, leads to construction rules and regulations that induce an innovation shift to more robust suspension bridges or new features that do not conflict with the regulations.

Henry Petroski, an American engineer specializing in failure analysis, has summarized these ideas in law of failures (Petroski, 1994): About 30 years after the introduction of a new type of bridge, a failure will occur and a bridge will collapse (Table 18.1).

After a failure, the rules change and the innovative activity shifts to new areas. A typical example is driving a car: The more experience a young car driver gains, the more risks he or she is prepared to take when driving. This is what the ancient Greeks called *hubris*, which indicates a loss of one's sense of reality and an overestimation of one's capabilities, especially when the person exhibiting it is in a position of power. Hubris was considered a crime in classical Athens, the biggest crime a person could commit, all contained in the tale of Daedalus and his son Icarus. Daedalus, the engineer in Greek mythology, is a typical expert who can solve every problem in the realm of his expertise but who does not recognize where his expertise ends. Today, hubris denotes an overconfident feeling of pride and arrogance; it is often associated with a lack of humility, though not always with a lack of knowledge. An accusation of hubris often implies that suffering or punishment will follow: "Pride goes before a fall". The downfall of the former Italian Prime Minister Silvio Berlusconi has, among other things, been attributed to his hubris in the international media. The mix of overconfidence, condescension and pride may have skewed his perception of the Italian economy in 2011, which ultimately led to his ousting. The term has also been applied to the attitudes of some corporations and governments toward the natural environment that led them to ignore the dangers of waste-water or oil spills. In the wake of the 2010 Deepwater Horizon oil spill, journalist Naomi Klein found that "the initial exploration plan that BP submitted to the US federal government for the ill-fated Deepwater Horizon

Table 18.1 Petroski's law: every 30 years a big bridge failure?

Dee	Trussed girder	Tensional instability	1847
Tay	Truss	Unstable in wind	1879
Quebec	Cantilever	Compressive buckling	1907
Tacoma	Suspension	Aerodynamic instability	1940
Milford	Box girder	Plate buckling	1970
?	Cable stayed?	Instability?	20xx?

well reads like a Greek tragedy about human hubris. The phrase 'little risk' appears five times" (Klein, 2010).

Schumpeter emphasized the destructive storms of innovation: something new does not come smoothly, and it has to destroy old and proven technology. In his time these destructive storms were not yet seen as global risks. Inventors and entrepreneurs like James Watt were cautious; they were fully aware of the insecurity surrounding their invention and did not completely trust their new technology. However, as in the days of James Watt, the next generation of engineers has more experience in the new technology, which leads to a change of behaviour: they are less cautious. On top of that, competition enforces cost reductions. The combination of being less cautious and trying to reduce costs can lead to disasters like the collapse of the bridge of Tay or the failure of the Hubble Telescope, where a cheap final test was cancelled. Watt himself never built engines with steam pressures much higher than atmospheric pressure because he was concerned that a boiler could explode, but his successors had no such qualms. Some decades later, Switzerland built the first tunnel through the Gotthard. Since coal-fired locomotives could not be used during its construction, the engineers used turbines driven by compressed air. Some of the air tanks exploded and dozens of people were killed.

Scientists take on risks not because they are careless, but because they feel the need to move into unknown fields. In general, scientists are not interested in probing what is already known (Fig. 18.2).

They want to detect something new, a surprise, which often means that they have to go beyond the frontiers of today's know-how.

Many of the early researchers in nuclear energy died of cancer; they were not aware of the dangers of nuclear radiation. Marie Curie's lab books are still radioactive today. Her papers from the 1890s are considered to be too dangerous to be handled like ordinary lab books. Even her cookbook is highly radioactive. They are kept in lead-lined boxes, and those who wish to consult them must wear protective clothing. Famous for her pioneering research on radioactivity, she was the first person to be honoured with two Nobel Prizes, one in physics and one in chemistry. She was the first female professor at the University of Paris, and in 1995 she became the first woman to be entombed on her own merits in the Panthéon in Paris (Fig. 18.3). She applied her vast knowledge about X rays in WW I helping wounded soldiers close to the front line with a mobile X ray station. Curie died in 1934 of aplastic anaemia, most probably brought on by her years of exposure to radiation.

Science is not the big danger. Researchers in laboratories work with very small doses of dangerous materials in strictly controlled environments. This is perhaps no

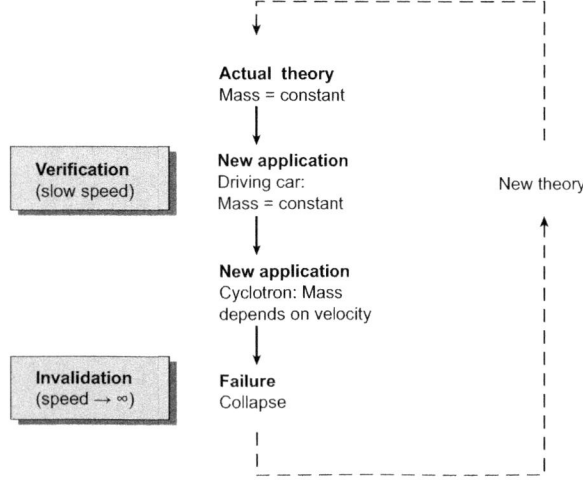

Fig. 18.2 Scientist deliberately seeks "surprise" or failure

longer valid for genetic experiments: Here, "living" objects are produced that can replicate themselves quickly. We depend on honesty and proper laboratory and manufacturing practices. Researchers who experiment with very dangerous virus strains, like the bird flu H5N1, work in special labs. Governments are cautious about publishing the results. A very dangerous strain of bird flu produced by researchers 2011 in Australia was only allowed to be published some years later. Big risks occur when new technology is moved to industrial scales because many risks do not increase linearly with scale but may grow much faster. One example is the heat increase in a chemical reaction which is proportional to volume and which increases at n^3. The surface that can be used to dissipate the heat increases at n^2. With size we add complexity and non-linear effects.

18.3 Perrow's Dilemma

Charles B. Perrow, an emeritus professor of sociology at Yale University, has summarized the most important dimensions of risk in modern technologies in his two dilemmas (Fig. 18.4). He came to these conclusions after a thorough investigation of the Three Mile Island accident in the US.

- 1st dilemma: Some technologies are complex and tightly coupled. Complex systems consist of many components and these components have close relationships as in a nuclear power station. Complexity needs decentralized decision-making, otherwise speed is lost in an accident and the only people who have expertise on the subject matter are the local experts. No single individual understands all of the modules of a nuclear power station. On the other hand, tight coupling needs centralized decision-making. Only a central office knows what happens to the whole system if one detail is changed.

Fig. 18.3 Leading academics: careless or heroic? Marie Curie (third from the left, first row) with Albert Einstein (fifth from the left, first row) and others at Brussels 1927

- 2nd dilemma: Complexity and tight coupling lead to systematic risks which can be overcome by backups. But backups increase complexity. And we are back to dilemma 1.

One example of loosely coupled systems with simple technological interactions is a traditional manufacturing company (Fig. 18.5). An example of a more complex technological system is the railway sector. It is tightly coupled. Loose coupling, but non-linear interactions can be found in the military or universities. Both have intense human interactions; they are adaptive. We know from chaos theory that such systems can quickly get out of control as well and prediction is impossible. Complex and tightly coupled systems include space missions or nuclear power stations where the systemic risks are high and need tight control.

Thus out of his dilemmas Perrow concludes: High technology is dangerous and there are no full technical fixes. Society has to live with systemic risks. When a risk becomes a political issue, a decision has to be made as to what kind of risks should be accepted and what kind of risks should be avoided. The human perception of risk

18.3 Perrow's Dilemma

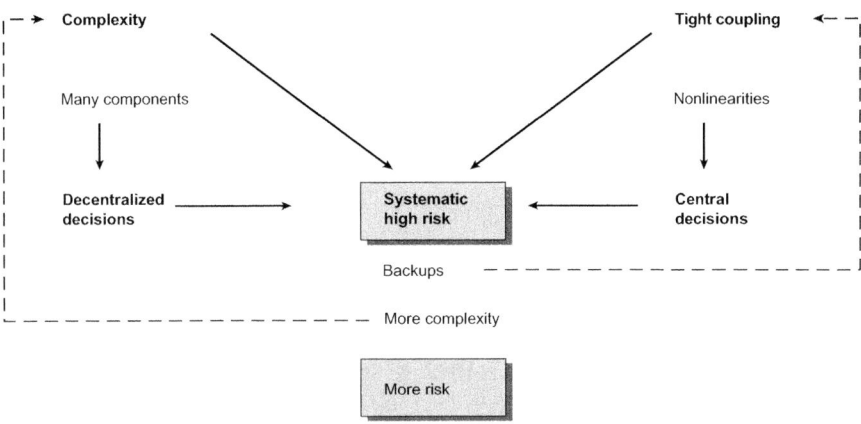

Fig. 18.4 Perrow's Dilemma: risk cannot be overcome by technology alone (Perrow, 1984)

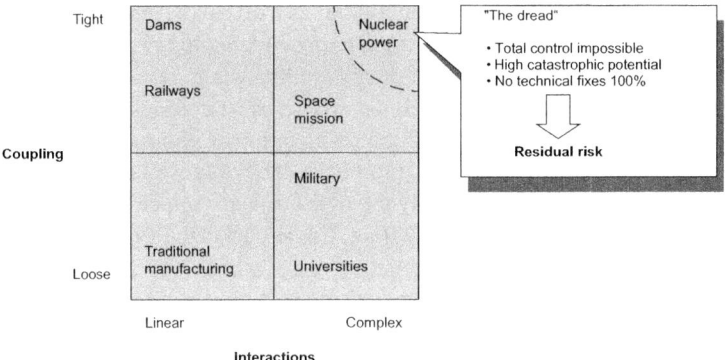

Fig. 18.5 Two dimensions of risk: coupling and interaction

becomes as important as the technical view. However, learning about the risk potential of a new technology takes time and thus increases costs.

The two dilemmas of Perrow also show what to do if we want to reduce risks in a system: We have to reduce complexity and tight coupling. This runs contrary to Just in Time concepts where we reduce all buffers between the different stages in a supply chain. We increase coupling. Many companies learnt these risks the hard way after Fukushima when Japanese suppliers could not deliver anymore.

Kahnemann and Tversky have studied the human perception of risk for more than 50 years. Kahneman established a cognitive basis for common human errors using heuristics and biases, and developed prospect theory. He was awarded the 2002 Nobel Prize in Economics for his work. In 2011 his book Thinking Fast and Slow, which summarizes much of his research, was published and became a best-seller (Kahneman, 2011). Kahnemann and Tversky (1979) have shown that risks are not judged rationally at all when probabilities are to be estimated. Probabilities are very useful whenever we have to set priorities for the allocation of scarce resources like safety improvements for an already safe railway. Probabilities are of little value when we have to make a judgement about the acceptance of risk in society.

18.4 Risk Acceptance

In democracies, governments have tried to institutionalize risk discussions in order to increase the general population's safety and to gain approval for new technologies: The US introduced technology assessment in 1973, Germany followed in 1990 and Switzerland in 1992. Researchers and experts try to estimate the effects of new technologies. Not everybody agrees with this approach. But every experienced engineer knows that technology forecasts and statistics can be wrong, especially if they concern the future. This is known as the *Collingridge dilemma* (Collingridge, 1980): on the one hand, the impact of new technologies cannot be easily predicted until the technology is extensively developed and widely used; on the other hand, controlling or changing a technology is difficult as soon as it is widely used. Technologies are often not applied where researchers or inventors think they are most useful. They develop unexpected trajectories when technologies are still in their infancy. It is like debugging a computer program; you never know in advance which applications will be run on it. The only way out is to conduct as many trials as possible and then apply the precautionary principle: That which you do not understand you should treat with extreme care. Don't eat mushrooms you cannot name.

A second approach, to convince the public that new technologies can be fully controlled, does not work. Some experts, especially researchers, believe that the public would support new technologies if only they were better educated. Statistics tell a different story. The supporters and opponents of new technologies are evenly distributed within academics and non-academics. Whether people oppose technologies or not depends mostly on their general acceptance of economic rationales, their liberalism and their trust in managers and administrators who run dangerous facilities. The main public concerns are:
1. Is the decision process acceptable to those who could be hurt?
2. Is there an agreement as to who will be held responsible for what?
3. Are there institutions that manage and regulate the technology in a trustworthy manner?

Engineers and insurance companies calculate the risk based on the number of deaths, illnesses, or property damage and the formula $Risk = probability * impact$. Society evaluates risks through political discussions.

When we follow discussions about nuclear waste disposals, we have discussions on three levels: The type of installation, the location and the technology itself. Whenever a technology is not accepted at all, it makes no sense to discuss where it should be placed, and as long as the location is not accepted, it is useless to talk about the specific type of installation. In order for a technology to be accepted, it is very important that people feel as if they themselves have made a decision based on their free will. For instance, nuclear medicine is not a controversial issue.

Safety is a general guiding principle in developed countries. The richer a country is the more precision it expects from its forecasts. Emotions produce expectations which have to be taken seriously in risk assessments. This leads to measures that increase safety and create a climate of trust. For example, the "Hazard analysis and critical control points" (HACCP) is a systematic preventive approach to food and pharmaceutical safety that identifies physical, chemical, and biological hazards in the production processes that can cause the finished product to be unsafe. It provides measures for reducing these risks to a safe level. HACCP is referred to as a hazard prevention protocol rather than a product inspection. Originally, NASA introduced it in 1960 in order to ensure the food quality for space flights. Today it is field-tested and has an established routine. The seven HACCP principles are included in the international standard ISO 22000 FSMS 2005.
1. Conduct hazard analysis
2. Identify critical control points
3. Establish limits for each point
4. Establish monitoring requirements
5. Establish corrective actions
6. Establish record keeping
7. Monitor whether HACCP is working

For high reliability organizations (HROs) such as nuclear power plants, oil companies or hospitals, an appropriate risk and safety management is even more important. For example, in the past, the Swiss Railway Company Primary Control Unit separated safety management from efficiency: One central group of five engineers had the complete overview of one third of the system. The group was responsible for efficiency: Where does a fast passenger train overtake a slow freight train? In Switzerland freight trains and passenger trains use the same tracks. But safety was fully in the hands of the stationmasters: They switched signals from red to green. The central group could not influence the system directly. Even if the central group failed in their duty, safety was guaranteed. Whenever efficiency and safety have similar levels of priority and both tasks are combined in one person, risks may get out of control. Air traffic controllers live with this dilemma each day but they are trained for safety first.

Trusting in the institutions that oversee, regulate or run risky technologies is the cornerstone of technology acceptance. Modern society needs high reliability organizations.

The Economist: Genetic modification: Filling tomorrow's rice bowl

Genetic engineers are applying their skills to tropical crops
© The Economist Newspaper Limited, London (Dec 7th 2006)

EVERY hectare of paddy fields in Asia provides enough rice to feed 27 people. Fifty years from now, according to some projections, each hectare will have to cater for 43. Converting more land to paddy is not an option, since suitable plots are already in short supply. In fact, in many of the continent's most fertile river basins, urban sprawl is consuming growing quantities of prime rice-farming land. Moreover, global warming is likely to make farmers' lives increasingly difficult, by causing more frequent droughts in some places and worse flooding in others. Scientists at the International Rice Research Institute (IRRI) doubt it is possible to improve productivity as much as is needed through better farming practices or the adoption of new strains derived from conventional cross-breeding. Instead, they aim to improve rice yields by 50% using modern genetic techniques.

On December 4th the Consultative Group on International Agricultural Research (CGIAR), a network of research institutes of which IRRI is a member, unveiled a series of schemes intended to protect crop yields against the ill effects of global warming. Many involve genetic engineering—which is generally embraced by farmers in poor countries even if some Western consumers turn their noses up at it. Some, though, only use genetics to identify useful genes. For example, IRRI's scientists have found a gene that allows an Indian rice strain to survive total immersion for several weeks, and have cross-bred it into a strain favoured by farmers in flood-prone Bangladesh. In trials, the new plant produced as much rice as the original under normal conditions, but over twice as much after prolonged flooding. This trait could increase the world's rice harvest dramatically, since flooding damages some 20m hectares (50m acres) of rice each year out of a total crop of 150m hectares.

By far the most ambitious project on CGIAR's list, though, involves transforming the way in which rice photosynthesises. That will require some serious genetic restructuring.

Three into four will go

Most plants use an enzyme called rubisco to convert carbon dioxide (CO_2) into sugars containing three carbon atoms—a process known as C3 photosynthesis. But at temperatures above 25°C, rubisco begins to bond with oxygen instead of CO_2, reducing the efficiency of the reaction. As a result, certain plants in warm climates have evolved a different mechanism, called C4 photosynthesis, in which other enzymes help to concentrate CO_2 around the rubisco, and the initial result is a four-carbon sugar. In hot, sunny climes, these C4 plants are half as efficient again as their C3 counterparts. They also use less water and nitrogen. The result, in the case of staple crops, is higher yields in tougher conditions: a hectare of rice, a C3 plant, produces a harvest of no more than eight tonnes, whereas maize, a C4 plant, yields as much as 12 tonnes.

Turning a C3 plant into a C4 one, though, is trickier than conferring flood resistance, since it involves wholesale changes in anatomy. C4 plants often absorb CO_2 from the air in one type of cell and then convert it to sugars through photosynthesis in another. C3 plants, by contrast, do both jobs in the same place.

On the other hand, C4 photosynthesis seems to have evolved more than 50 times, in 19 families of plant. That variety suggests the shift from one form of photosynthesis to the other is not as radical as might appear at first sight. It also gives researchers a number of starting points for the project. Some C4 plants, for example, absorb CO_2 and photosynthesise it at either end of special elongated cells, instead of separating the functions out into two different types of cell. Many C3 plants, meanwhile, have several of the genes needed for C4 photosynthesis, but do not use them in the same way. In fact, the distinction between C3 and C4 plants is not always clear-cut. Some species use one method in their leaves and the other in their stems.

John Sheehy, one of IRRI's crop scientists, plans to screen the institute's collection of 6,000 varieties of wild rice to see if any of them display a predisposition for C4 photosynthesis. Other researchers, meanwhile, are trying to isolate the genes responsible for C4 plants' unusual anatomy and biochemistry. A few years ago, geneticists managed to get rice to produce one of the enzymes needed for C4 photosynthesis by transplanting the relevant gene from maize.

The task, admits Robert Zeigler, IRRI's director, is daunting, and will take ten years or more. But the potential is enormous. Success would not only increase yields, but also reduce the need for water and fertilisers, since C4 plants make more efficient use of both. Other important C3 crops, such as wheat, sweet potatoes and cassava, could also benefit. If it all works, a second green revolution beckons.

References

Collingridge, D. (1980). *The social control of technology*. London, UK: Pinter.
Kahneman, D. (2011). *Thinking fast and slow*. New York, NY: Farrar Strauss & Giroux.
Kahneman, D., & Tversky, A. (1979). Prospect theory: An analysis of decision under risk. *Econometrica: Journal of the Econometric Society, 47*, 263–291.
Klein, N. (2010, June 18). Gulf oil spill: A hole in the world. *The Guardian*, p. 19.
Perrow, C. (1984). *Normal accidents: Living with high risk systems*. New York, NY: Basic Books.
Petroski, H. (1994). *Design paradigms: Case histories of error and judgement in engineering*. Cambridge, MA: Cambridge University Press.

Non-proliferation of Adverse Technologies 19

> *Challenging the integrity of the non-proliferation regime is a matter which can affect international peace and security.*
> Mohamed ElBaradei, Egyptian diplomat and Nobel Peace Prize winner 2005

Abstract

In order to control the most powerful military technology, the atomic bomb, 190 states have signed the Non-Proliferation of Nuclear Weapons (NPT) treaty to date. Other treaties try to reduce the number of nuclear warheads and the amount of technology transfer to rogue states in general. So far transfer restrictions have worked well. Every company has to develop its own position in this critical situation. Modern technology is ever more powerful and cannot be kept secret. The threat for companies and for society being taken by surprise is increasing.

Technology is said to have two origins, canons and watches, either used for good or for bad, for peace or for war. Remember Archimedes, one of the leading scientists in antiquity: With his great technical knowledge, he helped Syracuse, his hometown, to defend itself with all his technical tricks. That is why the Roman conquerors gave the order not to harm Archimedes. However, while contemplating a mathematical diagram, the city was captured and a Roman soldier forced him to stop his work, but Archimedes declined, arguing that he had to finish the problem. Uttering his last sentence: "Do not destroy my circles", Archimedes lost his life.

Wars have always had a great influence on technology. In the twentieth century wars also had a huge impact on research activities. Up until World War I, rich individuals and corporations funded research to a great extent, but the new technology that was used in World War I and even more in World War II proved to be very powerful. For this reason, governments like Germany, France, Great Britain and the US started to invest heavily in research activities to develop sophisticated weapons. Aeroplanes, submarines, radar, toxic gases, calculations of ballistic trajectories and optics made big improvements due to scientific research. In fact, the two world wars triggered great science which can be both beneficial and very dangerous. It needs to be balanced by governments and international contracts (Fig. 19.1).

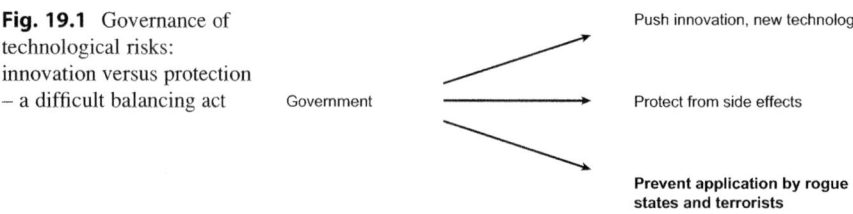

Fig. 19.1 Governance of technological risks: innovation versus protection – a difficult balancing act

On the one hand, innovations and new technologies have to be supported in order to grow. On the other hand, the side-effects of these developments have to be minimized and rogue states and terrorists should not be able to gain access to the most powerful technologies. That is why the Nuclear Non-Proliferation Treaty (NPT) came into force on March 5th, 1970 and was extended indefinitely in 1995. It has three pillars: Non-proliferation, disarmament and peaceful use of nuclear energy. The NPT is the most widely accepted arms control agreement. Only Israel, India and Pakistan have never signed the Treaty. North Korea withdrew from the treaty in 2003.

19.1 Nuclear Power

Nuclear power goes back to Becquerel's discovery in 1896 that uranium emits radiation. In 1903 Rutherford already saw the possibility of inexhaustible power, white power. But it took another 35 years to understand the mechanism of a nuclear chain reaction. The US started one of the biggest projects ever undertaken, which resulted in the destruction of Hiroshima and Nagasaki.

After the war the scientific community felt somehow betrayed and wanted to implement a peaceful application of nuclear technology. President Dwight Eisenhower started his presidential program "Atoms for Peace" in 1953, emphasizing the need to quickly develop "peaceful" uses of nuclear power. The US industry ordered 218 power stations after General Electric's first privately funded nuclear power reactor provided electricity for the state of California. The government covered insurance for the worst-case scenario. But taming the new technology proved to be far more difficult than anticipated. Concerns increased and the program was finally cancelled, not by its adversaries, but because of the heavy burden the safety precautions placed on it. Under President Carter about 100 orders were cancelled with an estimated loss of USD 74 billion for the US industry. Since then, the industry knows that it is important to monitor new technology carefully, regardless of whether it will be accepted or not; and if it is accepted, under which rules and standards. Laws are less predictable than technology roadmaps and may change very fast as we now know since Fukushima.

Nuclear power had to be tamed both from a technical and a political standpoint: A whole set of international treaties were drafted, among them the NPT. The NPT is an international treaty with the goal of preventing the proliferation of nuclear weapons and weapons technology, of promoting the co-operation between countries in the peaceful uses of nuclear energy and of reaching the goal of nuclear disarmament. It was signed by more than 190 nations, including the five countries that officially have nuclear weapons: the US, Russia, UK, France and China. Four countries that have not signed the treaty are known or are believed to possess nuclear weapons: India, Pakistan and North Korea have openly tested or have declared that they possess nuclear weapons and Israel remains secretive, refusing to disclose any information concerning its own nuclear weapons program.

A variety of treaties and agreements have been enacted to regulate the use, development and possession of the various types of weapons of mass destruction. Treaties may regulate weapons' use under the customs of war (Hague Conventions, Geneva Protocol), ban specific types of weapons (Chemical Weapons Convention, Biological Weapons Convention), limit weapons research (Partial Test Ban Treaty, Comprehensive Nuclear-Test-Ban Treaty), limit allowable weapons stockpiles and delivery systems (START I, SORT) or regulate civilian use of weapon precursors (Chemical Weapons Convention, Biological Weapons Convention). The history of weapons control includes treaties to limit the effective defence against weapons of mass destruction in order to preserve the deterrent doctrine of mutually assured destruction (Anti-Ballistic Missile Treaty). We also have treaties to limit the spread of nuclear technologies geographically (African Nuclear Weapons Free Zone Treaty, Nuclear Non-Proliferation Treaty). Many ideas behind these treaties rely on modern game theory: Defence is symmetrical to aggression. Thus defence has to be controlled as well.

19.2 Technology and Defence

Napoleon Bonaparte once boasted he could gamble with two million lives per year in wars. In his time we saw the first big armies in human history, which would not have been possible without the manufacturing technologies of the first industrial revolution. Most historians nowadays agree that this revolution was not James Watt's steam power, a technology, but rather the American system of mass production, an organizational innovation. Only after this new production method had been invented by Whitney for textile machines was it possible to arm a levee en masse, to mobilize large numbers of soldiers. Before 1790 weapons were produced piece by piece in small sweatshops with a very low productivity. The first industrial revolution made use of specialization and could harvest the fruits of economy of scale.

Today the US spends about USD 700 billion on its military; the 2010 budget was USD 664 billion, which is 50 % of the total global expenditures. USD 80 billion go into military research (Table 19.1). This is about 11 times more than Switzerland spends on R&D in total. But civilian and military R&D converge more and more. The same is true for technologies: Global positioning systems, at the beginning a military system, are nowadays used for more civilian applications. Thus we have more and more double use goods: Products that can be used for both, military and civilian application.

Table 19.1 Defence technology: huge consumption of resources, particularly in the US

Expenditures for defence	World 2011	USD 1,630 bn
	US	USD 711 bn
	China	USD 143 bn
	UK, F, Japan	~USD 60 bn
US expenditures for military R&D	1980	USD 14 bn
	1990	USD 40 bn
	1997	USD 33 bn
	2010	USD 79 bn

19.3 NATO's Priorities

Not everybody can spend as much as the US on weapons. Since the fall of the Berlin Wall in 1989, most countries' priority has been to reduce costs for their armies. They have to focus onto the most important threats.

For more than 15 years the North Atlantic Treaty Organization's (NATO) priority list has not changed: Defusing proliferation incentives has always been on top of the list.
1. Defusing proliferation incentives
2. Enforcing international sanctions
3. Offensive military actions
4. Ballistic missile defence

In order to prevent proliferation, NATO has always wanted to keep some technologies under tight control. This includes nuclear technology as well as missile technology that can be used to transport nuclear weapons. Two treaties were imposed, the nuclear non-proliferation treaty (NPT) and the missile technology control regime (MTCR) (Fig. 19.2). The MTCR is an informal and voluntary partnership among 34 countries with the goal of preventing the proliferation of missile and unmanned aerial vehicle technology capable of carrying a 500 kg payload for at least 300 km.

But NATO's attitude has been blurred from the beginning. It often seems that it adheres to the "the enemy of my enemy is my friend" philosophy. Thus it indirectly supported Israel's efforts to develop a bomb. Although Israel has most probably already been in possession of a nuclear weapon since the late 1960s, it has never publicly confirmed its nuclear weapons program. But Mordechai Vanunu, a former Israeli nuclear technician, revealed details of the program to the British press in 1986. Israel has never officially admitted to having nuclear weapons; it has always repeated that it would not be the first country to "introduce" nuclear weapons to the Middle East. This statement does not give us any information: it is unclear whether Israel will not create, will not disclose or will not make the first use of the weapons. Israel has refused to sign the NPT despite international pressure to do so, and has stated that signing the NPT would be contrary to its national security interests. In the Vela incident on September 22nd 1979, the US satellite Vela probably detected

19.3 NATO's Priorities

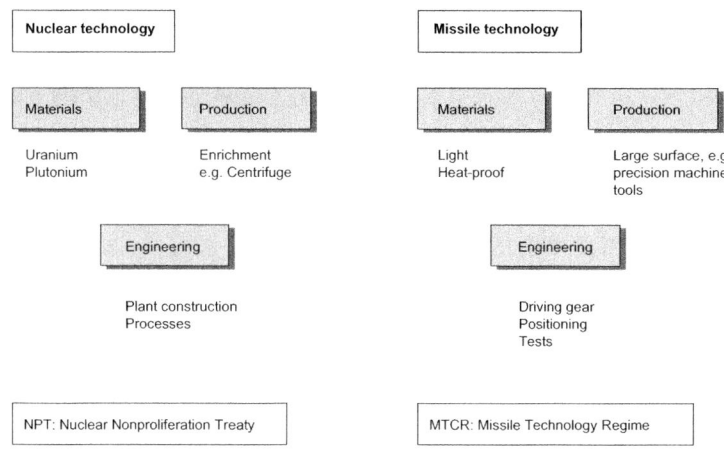

Fig. 19.2 Top priority: nuclear technology and missiles

nuclear tests in the Indian Ocean, as flashes of light were detected which resembled a nuclear detonation. The official statement was that the flashes were just a result of natural occurrences. Nonetheless, some journalists believe that this was the third joint Israeli-South African nuclear test. Today, Israel is believed to have several hundred nuclear warheads at its disposal and to be able to transport them on intercontinental ballistic missiles, aircrafts and submarines.[1] The estimates on how many warheads Israel has built since the late 1960s vary and are mainly based on the amount of fissile material that could have been produced and on the revelations of Israeli nuclear technician Mordechai Vanunu. The costs of the total program have been estimated at about USD 10 billion.

Alone or with other nations, Israel has used diplomatic and military efforts as well as covert actions to prevent other Middle Eastern countries from acquiring nuclear weapons. Arab nations have accused the US of practising a double standard in criticizing Iran's nuclear program while ignoring Israel's possession of nuclear weapons.

Contrary to Israel's efforts, the Iraqi nuclear program was not a success. Iraq also spent more than USD 10 billion on its program, but Israel bombed the Ozirak reactor in 1981 while the secret service Mossad ran a covert operation to infiltrate the Iraqi nuclear program and removed some associated personalities after the air strike. The Iraqi nuclear program had to use relatively inefficient methods of uranium enrichment. They actually had to go back to the same method the Americans used in their Manhattan project in World War II. The non-proliferation efforts were successful in that the program was delayed for several years. In 1991, at the outbreak of the first Gulf war, Iraq was still some years away from completing its project and would have been capable of building a

---

[1] Israel – Nuclear Weapons, Federation of American Scientists. Retrieved 1 July 2007.

nuclear bomb in 2–4 years. After the second Gulf war no weapons of mass destruction were found: non-proliferation was successful again.

19.4 Adverse Technologies

Defence Technology needs four pools of know-how:
1. Systems have to be engineered. Somebody has to understand the total system with all its technical interfaces and modules.
2. Weapon systems have to be produced. Usually you need specific machine tools. Silent submarines need the most robust 5 axis milling machines to deliver the tight tolerances needed for driving the ship's propellers. Starrag, a Swiss machine producer, was not allowed to export its machines to the Eastern bloc countries during the Cold War. Stuxnet, a computer worm, probably disabled Iranian centrifuges. Stuxnet invaded some 60 % of all Iranian computers, especially those that were running Siemens controllers.
3. For nuclear applications you need special materials such as plutonium and uranium.
4. In the end, weapons have to be deployed. This needs rocket know-how.

All this expertise has to be transferred or to be developed from scratch. Thus there are two approaches to prevent proliferation:
1. Either you declare a list of dangerous materials, machines or technologies and control the possession of these technologies,
2. Or you have a list of states and make sure the black-listed ones do not get access to the top technologies.

Proliferation or non-proliferation is a fundamental decision for every corporation: A company has to decide between a policy of opportunity and a policy of commitment (Fig. 19.3).

In any case, management has to allocate responsibility and take all the possibilities of infringement and cannot delegate responsibility to lower ranks. In most cases companies have to issue an end-user certificate to make sure where the technology will actually be installed. There are US rules about the content of products as well. Since 2004 products which contain more than 20 % of American value added are not allowed to be delivered to the "Axis of Evil". Former US president George W. Bush coined this term in 2002, a few months after the September 2001 attack on the World Trade Centre in New York. Bush claimed that Iran, Iraq and North Korea supported terrorism and had developed weapons of mass destruction. However, the percentage of US value added is difficult to quantify since supply chains have become very complicated.

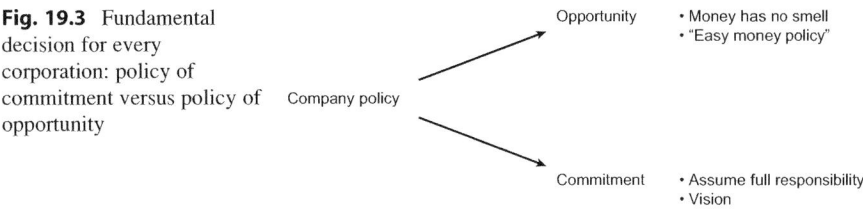

Fig. 19.3 Fundamental decision for every corporation: policy of commitment versus policy of opportunity

19.5 International Treaties

In 1949 the West established the CoCom, the Coordinating Committee for Multilateral Export Controls, to ensure the technological lead of western nations. It worked well but was superseded by the Wassenaar arrangement in 1995. It lists potentially dangerous technologies and dual use goods, which are goods that have both military and civilian use. The list is updated every year and all of the participating countries give each other updates on a regular basis. The basic list is composed of ten categories based on increasing levels of sophistication. Some categories are, for example, Special Materials and Related Equipment, Materials Processing, Electronics, Computers, Marine, Aerospace etc.

Despite all of the treaties, lists and arrangements, it has to be kept in mind that there are no evil technologies; there are only malicious people who misuse these technologies. Thus black lists will be with us for many years to come. As good corporate citizens we have to be aware of this danger and have to support the control of transfer of both dual use and single use goods.

19.6 Terror and Epidemics

Terror is cheap and modern society is vulnerable: The cost of conducting a terror attack, such as 9/11 in New York, or the London and Madrid bombings, is very low, especially when seen in relation to the financial impact and the number of fatalities. Terrorist attacks are by far cheaper than big wars. For this reason they cannot be brought under control by checking streams of finances only.

This is one of the reasons why the US government has been worrying about further terror attacks since 2001. The US authorities are concerned about biological attacks. But these are carried out in a very different way. None of the past biological and chemical attacks have been successful so far: in the US, Anthrax killed five people, even though several hundred people were exposed to the bacteria. There was no SARS or smallpox epidemic and neither was any biological program found in Iraq.

Natural germs can be very dangerous. In history, smallpox killed more than 500 million people in less than 50 years.

However, there will be no Smallpox epidemic for three reasons:
1. Access to bacteria is very limited

2. It is easy to diagnose
3. Vaccination is effective even after infection

Nevertheless, even though countries like Switzerland have trained their politicians and military forces to deal with a pandemic horror scenario with hundreds of thousands of casualties, the problem of not being able to stock vaccines remains. The industrial structure of Western countries makes it difficult to produce a lot of vaccines on short notice because production has been outsourced due to specialization and labour costs. Travel would lead to a fast diffusion of the bacteria. By using databases and models, American researchers from Harvard University predicted that there would be one pandemic every 40–50 years. However, these statistics do not help to prevent epidemics. We do not need sophisticated databases; we need professional and standardized disease control measures. These are already in place in most developed countries and are ready to defend us against any natural or artificial epidemics. Uncertainty remains. Nobody knows how long it takes to develop a new vaccine.

The Economist: Nuclear security: Threat multiplier

Dangerous complacency about nuclear terrorism
© The Economist Newspaper Limited, London (Mar 31st 2012)

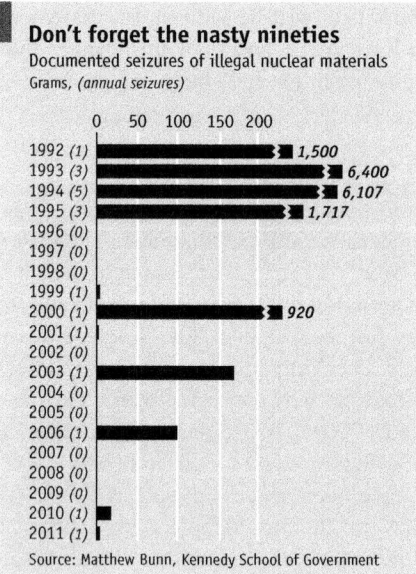

NUCLEAR mayhem can come from rogue states or badly run power stations. That fact escapes nobody in South Korea, just a mountain range away from rocket-mad North Korea, and with Japan's stricken Fukushima Dai-ichi plant across the sea. But it was a third threat—terrorism—that brought leaders from 53 countries to a summit in Seoul on March 26th-27th. It marked the halfway point in Barack Obama's four-year initiative to secure and reduce the world's scattered stocks of bomb-usable plutonium and highly enriched uranium (HEU). Though attendance was strong, the momentum shows worrying signs of slowing.

This is partly because the terrorist threat has waned since the killing of Osama bin Laden. That disrupted al-Qaeda, which under his leadership sought to obtain nuclear materials. A crackdown on black markets has been a big success. In the 1990s seizures were frequent and measured in kilos. The latest have mostly involved mere grams (see chart).

Now the easy targets have been reached. In the past two years, eight countries have disposed of some 480kg of HEU. Ukraine and Mexico have given up all their stocks. Kazakhstan has sealed away 13 tonnes of HEU and weapons-grade plutonium. A few countries have converted research reactors away from HEU. Belgium, France and the Netherlands have cut the amount they make for medical isotopes.

But the summit in Seoul called only for further voluntary reductions by the end of 2013. That is a weak commitment: the International Panel on Fissile Material estimates world stocks of HEU at 1,300 tonnes, plus 450 tonnes of separated plutonium. Although most of this is held by America and Russia (which counts as safe), the rest is scattered throughout more than 30 countries, some of it—according to Matthew Bunn of Harvard's Kennedy School—overseen only by night watchmen behind chain-link fences. Nobody named names at the summit, but a study this year by the Washington-based Nuclear Threat Initiative and the Economist Intelligence Unit (our sister company) said China, India and Japan scored poorly on security, and Uzbekistan and Vietnam did worst.

Vulnerabilities to terrorism remain. The Fukushima accident made it dramatically clear that nuclear power stations, if they can be knocked out by natural disasters, can also be hit by man-made assault.

The main hurdle to progress is sovereignty. Pakistan rejects almost any outside interference with its nuclear stockpiles, which are increasing (and, in outsiders' view, poorly guarded). Other countries resent being told what to do. In Seoul America and Russia did little more than repeat their 2010 commitment to dispose of 68 tonnes of weapons-grade plutonium. They may need to lead more by example if they want others to follow.

Glossary

Adverse technology Technology that can be used to produce powerful weapons of mass destruction. It may be double use or single use.

Aesthetics A set of principles that teaches how an object creates values, a perception of beauty, in our eyes: Understanding the whole without analysing the details.

Black list A list of traders, countries or technologies, with which a business is not prepared to trade, e.g. for political reasons (also denied list).

Bottleneck technology A technical module that limits the overall capacity of a system, e.g. the availability of charging stations limits the use of electrical cars.

Brand An identifying mark or characteristic common to the products of a particular company. Origin: brand = burning

Cannibalism Replacement of existing products before they run out of steam to make sure competitors cannot catch up.

Class action A lawsuit filed or defended by an individual acting on behalf of a group. In the US a lawyer represents a group of people that want to sue the same company for damages. The lawyer is paid a percentage of the compensation.

Co-opetition A combination of co-operation and competition that has been coined to describe cooperative competition. Samsung competes against Apple in the mobile phone market and is at the same time a major supplier of Apple.

Commercial law Laws that regulate property, contracts, taxes, etc. Everything that is needed to do business in a decent and efficient way. It is grounded in the general law of contracts.

Connectors People who know "everyone" and are able to change the perception of many potential customers.

Core capability A capability of an organization that was acquired in a long learning process, which has a positive impact on its perception through the customers and cannot be easily imitated.

Design Plan or drawing produced to show look, functions or workings of a product before it is built or made.

Dilemma A situation in which a difficult choice has to be made between alternatives that contradict each other, but both would be needed.

Diminishing returns An economic law: If an output depends on two parameters and one of the input parameters is fixed, the additional return diminishes with the increase of the other parameter. The output of a high performance computer depends on software and hardware. Improving hardware alone creates less and less additional problem solving efficiency.

Disruptive innovation Innovation that fulfils new customer needs and has the potential to substitute later existing products. It can be defined only in hindsight (There are different views on this point). Today, USB sticks can replace CDs at much lower cost per megabyte. But they were much more expensive when they were introduced on the market.

Dominant design A design that becomes an industrial standard for a specific product. As soon as it emerges, industry attains security and is able to invest heavily and on a long-term basis. A consolidation may be the effect.

Double loop learning Reflective learning is learning how to change the governing variables where the goals may be sacrificed. It involves making value judgements and decisions about goals.

Dual use goods Goods that can be used both for civilian and military applications. Examples are binoculars or shotguns.

End-user certificate A certificate that specifies the ultimate user of a technology. The buyers have to guarantee that they do not hand over the technology to a third party without telling the original manufacturer.

Entry barriers Obstacles that prevent a company from starting a new activity, e.g. selling a new product.

Externality A consequence of an economic activity which affects third parties without being reflected in the market price.

Fracking A synonym for hydraulic fracturing. Deep below the earth's surface a horizontal hole is drilled for several kilometres into rock that contains oil or gas. The horizontal tube is treated with small explosions. For several weeks, water mixed with sand and chemicals is pumped into the tube to create and wash out tiny fractures along which fluids such as gas or oil may migrate to the tube. It is a technology that was developed 50 years ago and improved greatly over the last decade in the US.

Growth Growth in economy usually means growth of GDP. Since the Club of Rome was founded, qualitative growth, such as quality of life, increase of knowledge or life expectancies, have become as important, but some cannot be measured easily.

HACCP Hazard analysis and critical control point. A manufacturing standard in the food and pharmaceutical industry to ensure quality.

High-tech An industry where companies differentiate their products through better technology. In statistical terms: A company that spends more than 10 % of sales in R&D or has more than 15 % engineers in their workforce.

Increasing return A good that increases in value with every additional buyer.

Industrial design Aesthetic shaping of an industrial product to convey a message, to make it different and increase acceptance or improve understanding of functionality. For software this may mean interaction design.

Innovation Something new the company can sell to a customer.

Innovation – radical Radical innovations aim to create fundamentally new businesses, products and processes and their combinations. They implicate dramatic change in technologies and introduce new paradigms.

Innovation – routine Routine innovation develops with small improvements in an already existing range of products, services, businesses or processes for already existing customers in already existing market segments.

Inside-out design Going from function to aesthetic appeal, e.g. the design of a spoon.

Junk patent A patent with no real technical background. It is set up to harass competition or to collect money from infringers.

Law of big numbers The bigger the sample size, the closer the relative frequency comes to theoretical probability and the closer the distribution comes to a Gaussian distribution.

Life cycle Sales as a function of time. Usually bell-shaped.

Lock-in-effect In some industries, technologies differ only slightly, but switching costs are high and attempts to substitute them fail. Then early market share makes the race: The first market leader sets the standard; the technology does not change anymore, even if it is not the best one The early product wins, just because it is available and proven. An example is the QWERTY keyboard.

Mainstream science Science driven by a generally accepted paradigm, accepted by a majority of scientists.

Market leader The company with the biggest market share usually measured in turnover. It has the advantage of scale and usually reputation as well.

Mechatronics A multidisciplinary field of engineering including mechanical engineering, electrical engineering, control engineering and computer engineering. Originally, mechatronics just included the combination of mechanics and electronics, hence the combination of mechanics and electronics; however, as technical systems have become more and more complex the word has been "updated" during recent years to include more technical areas.

Miniaturization Miniaturization is the creation of ever smaller components for mechanical, optical, and electronic products and devices. These items get easier to carry, easier to store, and much more convenient to use. In the end they may be cheaper as well because they need less material.

Modularity A system is modular if it is broken down into subsystems (modules). Mutual dependencies of these subsystems should be reduced to a minimum or to a few defined interfaces. The key of modularity is that all modules are independent and convertible.

MTBF Mean time between failures. The average time between two failures of a module or a system.

Open product A product that is integrated into a bigger system and is able to cope with different systems. Mobiles can be used in different standards and can use apps from different providers.

Outpacing Companies like Gillette are capable of outpacing their competitors with their speed. They are on the market before their competitors.

Outside-in design Going from overall aesthetic appeal to the placement of the different functions, e.g. a perfume bottle.

Paradigm "A paradigm is what members of a scientific community, and they alone, share. Conversely, it is their possession of a common paradigm that constitutes a scientific community of a group of otherwise disparate men" (Thomas Kuhn). A world view drawn on theories and methodologies of a scientific community. It defines how problems are solved and fixes interpretations of fundamental concepts. It is the basis for efficient scientific work. We have paradigms in R&D as well.

Patent A government licence to an individual or a body conferring a right or title for a period, the sole right to make, use, or sell a technical invention.

Peer-to-peer (P2P) A technology that facilitates access to data stored on individual computers in a network. The server has only the addresses and directories and manages the access for upload and download.

Person skilled in the art (expert in patents) The imaginary person in patent rights that understands today's technology and is able to judge whether the invention proposed in the patent is inventive and novel.

Petty patent The German "Gebrauchsmuster". Very similar to the patent, but usually with a shorter validity and less stringent approval requirements.

Pioneer A company that enters the market as the first one selling in a new product category.

Pirate A person or organization that appropriates or reproduces work of another for profit without permission. Somebody or some organization that takes advantage of gaps in existing laws, created through new technologies.

Precautionary principle (Rio, Earth Summit 1992, principle 15) "In order to protect the environment, the precautionary approach shall be widely applied by states according to their capabilities where there are threats of serious or irreversible damage. Lack of full scientific certainty shall not be used as a reason for postponing cost-effective measures to prevent environmental degradation." (A cautious political statement!)

Precompetitive research Research at an early stage where the results cannot yet be directly applied to products. Companies are able to pool their forces without fearing anti-trust measures or direct competition.

Prior art Any written knowledge made available to the public before a patent was applied for.

Private property Things belonging to a particular person.

Product category Family of products that customers see as closely related. A company defines categories. Examples are different types of toilet paper, salads or drugs.

Project An undertaking in a company that runs outside of the organizational structure. It has a specific goal, defined resources and runs for a limited time. The project leader takes on some competencies that are usually with line managers only.

Proliferation of adverse technologies The delivery of technologies directly used in the production of weapons of mass destruction. Delivery of technologies which are put on a blacklist by the US government.

Reach Defines how many people can be reached by a message through a specific technology.

Reductionism The practice of analysing and describing a complex phenomenon in terms of its simple or fundamental constituents, especially when this is said to provide a sufficient explanation.

Reliability The probability that no defect occurs in a specific time interval, usually from zero to time t. Keep in mind that probabilities are defined for wide samples only!

Revolutionary science Science that is in contradiction to the existing scientific thinking. "Tradition-shattering complements to the tradition-bound activity of normal science." (Thomas Kuhn)

Richness The content of a message.

Risk The possibility that something might happen that makes it impossible or more difficult to reach a goal. Risks are often measured in financial terms: They are an expectation-value, a probability multiplied by its financial impact.

S-curve S-shaped curve, time on the horizontal axis, volume sold on the vertical axis. This is the schoolbook S-curve; reality is different.

Saddle The dump between early adopters and early majority where we have often a dip in sales that kills many start-ups. The dip in PCs was 30 % and went on for 7 years.

Science-driven Science-driven environments are uncertain and risky. Precise forecasts are not possible.

Simulation Simulation is the imitation of some real object, state of affairs, or process, with the help of physical models, carried out on computers. The weather is simulated to provide forecasts.

Single loop-learning A thermostat that detects when it is too cold or too hot and turns a cooling unit on or off has learnt to control the temperature of a room. It follows Deming's well-known Plan – Do – Check – Act process. Single loop learning improves efficiency, but it does change goals.

Spill-over effect Spill-over effects are externalities of an economic activity that affect those who are not directly involved.

Spin-off A company founded by members from a bigger company and a chunk of know-how from that company. The set-up is usually based on a mutual understanding; it is friendly.

Standardization Standardization is the process of developing and implementing technical norms or a model that can be used for comparative evaluations. It facilitates comparison and may lower costs with economy of scale.

Statistical bias Systematic distortion of a statistical result due to a factor not allowed for in the derivation of the result.

Storms of disruptive innovations A concept introduced by the Austrian economist Joseph Schumpeter. Innovations destroy old habits, old artefacts and lead to a disequilibrium. Equilibrium is the exception.

Sunk cost Cost that has already been incurred and cannot be recovered.

Technological change (growth) The change in technology that has an impact on the gross domestic product (GDP): It cannot be measured directly. It is a residual. For the period 1900–2003 technology was responsible for 1.7 % GDP growth p.a. in the US.

Technological progress Difficult to measure. Progress is always the reduction of a distance towards a goal. Thus the goal and the parameter measured have to be defined before progress can be measured. For example, ABS systems for cars were reduced from 16 kg to about 4 kg in 10 years, thus progress was 15 % per year. But the general notion of technological progress is a diffuse issue and is used in a very general sense in most media.

Technology From Greek technologia, describes systematic treatment, first used in the seventeenth Century.The application of scientific knowledge for practical purposes (Oxford dictionary, 2004)."Technology in its widest sense is the manipulation of nature for human material gain." Joel Mokyr, The Gifts of Athena)

Technology speed A flimsy concept, but widely used in management. It can be measured by the rate of technological change per time period, change in performance, cost, weight etc. A true measure is possible only if the goal and the changing parameter are defined. Then speed becomes the rate of approaching this goal and is of outmost importance for the management of technology.

Tinkering Attempting to repair or improve in a casual and non-systematic manner. Today seen as an important source of innovation. People who tinker around may have no well-defined goal, no specific theory, and no business plan. They just try and play around with what they are interested in. Surprise is what they are looking for.

Tipping point The point where sales start to grow fast. This is where the final breakthrough starts. It can be determined in hindsight only.

Tool Object used to help to perform a job. From old German meaning "prepare". For a long time, technology was perceived as a tool.

Trademark Sometimes interchangeably used with brand. Trademark is used as a wider term including signatures, packaging, colour, etc.

Value proposition The reason why the customer should buy the product.

Weapon grade Used for plutonium and enriched uranium. Grade means purity and thus suitability for the production of weapons. Examples are uranium 235 and plutonium 239.

Printed by Printforce, the Netherlands